HYDROPONICS MARIJUANA CULTIVATION FROM SCRATCH
Step - by - Step Practical Guide To Soilless Gardening.
Unveiling The Secrets of Green Revolution

WRITTEN BY

FORTUNE MILLS

COPYRIGHT © 2024 BY FORTUNE MILLS.
ALL RIGHTS RESERVED.
NO PORTION OF THIS BOOK MAY BE REPRODUCED IN ANY FORM WITHOUT WRITTEN PERMISSION FROM THE

PUBLISHER OR AUTHOR, EXCEPT AS PERMITTED BY THE U.S COPYRIGHT.

TABLE OF CONTENTS

TABLE OF CONTENTS 3
INTRODUCTION 7
   ADVANTAGES OF HYDROPONICS 8
   WHY CHOOSE HYDROPONIC FOR MARIJUANA
   CULTIVATION 10
CHAPTER 1 13
GETTING STARTED: SETTING UP YOUR HYDROPONIC
SYSTEM 13
   Types of Hydroponic Systems for Marijuana 14
   Choosing the Right System for Your Needs 16
   1.2 UNDERSTANDING NUTRIENTS 17
   Essential Nutrients for Marijuana Growth 17
   Nutrient Solutions and Mixtures 19
   Calculating and Adjusting Nutrient Levels 21
CHAPTER 2 23
SELECTING THE RIGHT STRAIN 23
   2:1 INTRODUCTION TO MARIJUANA STRAINS 23
   Indica vs. Sativa vs. Hybrid 25
   Understanding Cannabinoid Content 27
   Factors to consider when choosing a strain 29
   2.2 BEST STRAINS FOR HYDROPONIC CULTIVATION 31
   High Yield Strains 33
   Disease-Resistant Strains 35
   Fast growing strains 36
CHAPTER 3 39
ENVIRONMENT AND CLIMATE CONTROL 39

- 3.1 OPTIMIZING GROWING CONDITIONS 39
  - Temperature and Humidity Control 39
  - Light Requirements 41
  - CO2 Levels 42
- 3.2 MANAGING AIR CIRCULATION 44
  - Importance Of Airflow 44
  - Preventing Molds and Mildew 46
  - Ventilation Systems 48

CHAPTER 4 50
GERMINATION AND PROPAGATION 50
- 4.1 SEED SELECTION AND GERMINATION 50
  - Choosing Quality Seeds 50
  - Germination Techniques 52
  - Transplanting Seedlings 54
- 4.2 CLONING TECHNIQUES 56
  - Benefits Of Cloning 56
  - Methods Of Cloning 57
  - Rooting Hormones and Solutions 60

CHAPTER 5 62
VEGETATIVE STAGE: PROVIDING OPTIMAL CONDITIONS FOR GROWTH 62
- Light Cycles 62
- Nutrients Requirements 64
- Pruning and Training Techniques 66
- 5.1 PREVENTING PEST AND DISEASES 68
  - Common Pests 68
  - Common Diseases 69
  - Integrated Pest Management (IPM) Strategies 70
  - Organic Pest Control Methods 72

CHAPTER 6 75
FLOWERING STAGE 75
- 6.1 TRANSITIONING TO FLOWERING 75
  - Light Schedule Adjustments 75

- Nutrient Adjustments 77
- Environmental Considerations 79
- 6.2 MAXIMIZING FLOWERING AND BUD DEVELOPMENT 81
- Supplemental Lighting Techniques 81
- Flowering Boosters and Enhancers 83
- Harvest and Timing Techniques 86

CHAPTER 7 89
HARVESTING AND CURING 89
- 7:1 HARVESTING YOUR CROP 89
- Signs of Maturity 89
- Harvesting Techniques 91
- Trimming and Manicuring Techniques 93
- 7.2 CURING AND DRYING PROCESS 95
- Importance of Curing 95
- Proper Drying Techniques 97
- Curing Jars and Storage 99

CHAPTER 8 102
TROUBLESHOOTING COMMON ISSUES 102
- 8.1 NUTRIENT DEFICIENCIES AND EXCESSES 102
- Identifying Nutrient Imbalances 102
- Corrective Measures 104
- 8.2 PEST AND DISEASE MANAGEMENT 107
- Recognizing Symptoms 107
- Treatment Options 109

CHAPTER 9 112
ADVANCED TECHNIQUES AND INNOVATIONS 112
- 9.1 HYDROPONICS SYSTEMS INNOVATIONS 112
- Aeroponics 112
- Deep Water Culture 114
- Nutrient Film Technique (NFT) 116
- 9.2 HIGH-TECH GROWING EQUIPMENT 119
- Automated Systems 119

Environmental Sensors 121
Data Driving Cultivation 123
CHAPTER 10 126
LEGAL CONSIDERATIONS AND COMPLIANCE 126
   10.1 UNDERSTANDING MARIJUANA LAWS 126
   Federal VS State Regulations 126
   Licensing and Permit Requirements 128
   Compliance With Local Zoning Laws 130
   10.2 ENSURING SECURITY AND DISCRETION 132
   Security Measures for Cultivation Facilities 132
   Transporting and Selling Marijuana Legally 134
   Risk Management Strategies 136
CONCLUSION 139
   Final Thoughts On Hydroponic Marijuana Cultivation 139
   Looking Ahead: Future Trends and Development 140
APPENDIX 142
   GLOSSARY OF TERMS 142
   RESOURCES AND REFERENCES 144
   RECOMMENDED SUPPLIERS AND BRANDS 146
INDEX 147

INTRODUCTION

In the realm of cannabis cultivation, innovation is the name of the game. From clandestine basements to state-of-the-art facilities, the journey of marijuana cultivation has evolved dramatically over the years. And at the forefront of this green revolution lies a groundbreaking technique that's redefining the way we grow cannabis: hydroponics.

Imagine a world where soil is no longer a necessity for thriving marijuana plants. Picture gardens where water, nutrients, and light converge in perfect harmony, creating an optimal environment for lush, potent buds to flourish. This is the essence of hydroponic marijuana cultivation – a method that's revolutionizing the cannabis industry one harvest at a time.

But what exactly is hydroponics, and why has it captured the imagination of cannabis enthusiasts and cultivators alike? At its core, hydroponics is a soil-less cultivation technique that relies on water-based nutrient solutions to feed and nourish plants. By eliminating soil from the equation, hydroponic systems offer unparalleled control over growing conditions, allowing cultivators to fine-tune every aspect of the plant's environment for maximum growth and potency.

The allure of hydroponic cultivation lies not only in its efficiency but also in its ability to produce high-quality cannabis consistently. With traditional soil-based methods, cultivators often face challenges such as nutrient imbalances, soil-borne pests, and limited space for root expansion. Hydroponic systems, on the other hand, provide a clean, sterile environment where these issues are virtually non-existent, allowing plants to reach their full genetic potential without the hindrance of soil-related constraints.

But hydroponics is more than just a method – it's a philosophy that embodies the spirit of innovation and experimentation. From DIY setups crafted from household materials to cutting-edge commercial facilities equipped with state-of-the-art technology, hydroponic cultivators are constantly pushing the boundaries of what's possible in the world of cannabis cultivation.

In this comprehensive guide, we'll delve deep into the world of hydroponic marijuana cultivation, exploring everything from the fundamentals of hydroponic systems to advanced techniques for maximizing yield and potency. Whether you're a seasoned cultivator looking to take your skills to the next level or a novice eager to embark on your first hydroponic adventure, this book is your road map to success in the world of soilless cannabis cultivation.

Throughout these pages, you'll discover the secrets of selecting the right strains for hydroponic growth, optimizing environmental conditions for maximum plant health and vitality, and navigating the complexities of nutrient management with confidence. You'll learn how to propagate plants from seed or clone, guide them through the vegetative and flowering stages with precision, and harvest the fruits of your labor with pride.

But beyond the practicalities of cultivation, this book is also a celebration of the passion and dedication that drive hydroponic cultivators to push the boundaries of what's possible. It's a testament to the ingenuity and creativity of a community united by a shared love for the plant and a relentless pursuit of excellence.

So join us on this journey into the heart of hydroponic marijuana cultivation, where innovation meets tradition, and the future of cannabis cultivation is yours to explore. Whether you're a hobbyist, a commercial cultivator, or simply a curious observer, the world of hydroponic cannabis cultivation awaits – and the possibilities are endless

ADVANTAGES OF HYDROPONICS

With so many advantages over conventional soil-based approaches, hydroponic marijuana production is becoming the method of choice for many growers. Hydroponics is revolutionizing the cannabis cultivation industry with its increased production possibilities and careful monitoring of growth factors. The following is a summary of the main benefits of hydroponic marijuana cultivation:

1.**Increased Yield Potential:** One of hydroponic farming's main benefits is its ability to produce larger yields than traditional soil-based methods. In hydroponic systems, plants receive the ideal amounts of water and nutrients directly at their roots, which promotes rapid and vigorous growth. Additionally, producers can adapt these systems to maximize spatial use, allowing them to nurture more plants in less spaces and increasing overall productivity.

2.**Faster Growth Rates:** Plants grown in hydroponic systems receive essential nutrients right away, avoiding the need for roots to sift through dirt. Faster growth rates and shorter crop cycles resulting from this optimized nutrient uptake enable growers to harvest more often and increase total production.

3.**Resource Conservation:** Compared to their soil-based equivalents, hydroponic systems have a higher resource efficiency by design. In contrast to traditional gardening, where water and nutrients may drain into the soil or become unavailable to plants, minimal waste is produced by the recirculation of water and nutrients inside the system. Furthermore, because water is supplied directly to the roots in hydroponic systems, there is less chance of evaporation and runoff, resulting in an overall reduction in water requirements.

4.**Precise Control Over Growing Conditions:** Perhaps one of the most significant advantages of hydroponic cultivation is the ability to control and optimize growing conditions with precision. Hydroponic systems allow growers to adjust factors such as nutrient levels, pH balance, temperature, and humidity with ease, ensuring that plants receive exactly what they need to thrive. This level of control minimizes the risk of nutrient deficiencies or imbalances, leading to healthier, more resilient plants.

5. Decreased Susceptibility to Pests and Diseases: Traditional cannabis cultivation is severely hampered by soil-borne pests and diseases. On the other hand, soil is completely unnecessary in hydroponic systems, which reduces the possibility of insect infestations and soil-borne infections. In addition, hydroponic systems are simple to sterilize in between harvests, which lowers the risk of contamination and promotes a hygienic, disease-free growing environment.

6. Optimal Space Usage: Hydroponic systems provide unmatched versatility in comparison to traditional gardening techniques in terms of spatial usage. Plants can be grown in a variety of ways, such as vertical gardens, stacked trays, or aeroponic towers, with nutrients being supplied straight to the root system. Cultivators can enhance space efficiency and grow more plants in a smaller area because of this versatility.

7. Year-Round Cultivation: Hydroponic systems are not limited by seasonal fluctuations or environmental factors, in contrast to outdoor cultivation. By controlling parameters such as light cycles, temperature, and humidity indoors, growers can create ideal conditions for cannabis growth, enabling year-round cultivation and the harvesting of multiple crops annually.

In conclusion, hydroponic marijuana cultivation offers an array of advantages that appeal to both novice and seasoned growers. From heightened yield potential and accelerated growth rates to precise environmental control and reduced susceptibility to pests and diseases, hydroponic systems revolutionize cannabis cultivation, heralding a more sustainable, efficient, and productive future for the industry.

WHY CHOOSE HYDROPONIC FOR MARIJUANA CULTIVATION

There are several good reasons to cultivate marijuana hydroponically, which is why producers are choosing it more and more. In the end, hydroponic systems produce higher-quality cannabis because they offer a regulated environment that maximizes effectiveness, speeds up growth, and optimizes yield. The following are some of the main arguments in favor of hydroponic marijuana cultivation:

1. Enhanced Growth Efficiency: Hydroponic systems deliver nutrients directly to the plant's roots, bypassing the need for roots to search through soil for sustenance. This direct nutrient uptake significantly enhances growth efficiency, leading to faster development and shorter crop cycles. With hydroponics, growers can achieve higher yields in less time compared to traditional soil-based methods.

2. Accurate Nutrient Management: One of the key characteristics of hydroponics is its capacity to accurately regulate the amount and makeup of nutrients in the growing medium. Growers can customize fertilizer formulations to match the unique requirements of their cannabis plants at various growth stages, guaranteeing the best possible nutrition and encouraging rapid development. Plants that have this exact nutrient control are healthier and more robust since it reduces the possibility of deficiencies or excesses.

3. Water Efficiency: Compared to conventional soil-based techniques, hydroponic systems are naturally more water-efficient. Hydroponics minimizes water waste and lowers total water usage by recirculating water inside the system. Furthermore, water is sent straight to the roots, reducing runoff and evaporation. For cannabis growing in areas with limited water resources or stringent water policies, hydroponics provides a viable option.

4. Space Optimization: Growers with limited space or those seeking to maximize productivity in a compact footprint will find hydroponic systems perfect as they may be tailored to enhance space use. Cultivators can grow more plants in less space by using aeroponic towers, stacked trays, or vertical farming techniques, which boosts total production and profitability.

5. Disease Prevention: Soil-borne pests and diseases pose significant challenges to traditional cannabis cultivation. However, hydroponic systems eliminate soil entirely, reducing the risk of pest infestations and soil-borne pathogens. Additionally, hydroponic setups can be sterilized between crops, further minimizing the risk of contamination and ensuring a clean, disease-free growing environment.

6. Climate Control: Growers may precisely regulate environmental parameters including temperature, humidity, and light levels using hydroponic systems. This control removes the limitations caused by seasonal changes and allows for year-round cultivation regardless of external climate conditions. Cultivators may optimize conditions for cannabis development and get consistent results year-round with climate-controlled indoor growing settings.

In conclusion, farmers that are looking for increased yields, effective resource management, precise control, and a sustainable cultivation method will find several benefits in choosing hydroponics. Hydroponics is an appealing alternative for both new and seasoned cannabis growers who want to regularly produce high-quality cannabis because of its capacity to speed up growth, enhance nutrition, conserve water, avoid disease, maximize space, and manage climate conditions.

CHAPTER 1

GETTING STARTED: SETTING UP YOUR HYDROPONIC SYSTEM

To achieve ideal plant growth and productivity, hydroponic system setup for marijuana cultivation involves meticulous planning and attention to detail. Here's a step-by-step tutorial to assist you successfully set up your hydroponic system:

1. Select the Correct System: Hydroponic systems come in different forms, each with unique benefits and factors to take into account. Deep Water Culture (DWC), Nutrient Film Technique (NFT), Ebb and Flow (Flood and Drain), and Drip Systems are examples of common varieties. Find out which kind best fits your demands and growth environment by researching each one.

2. Choose a growth Medium: Although hydroponic systems do not require soil, a growth medium is still required to stabilize and maintain the roots of the plant. Clay pellets, coconut coir, perlite, vermiculite, and rockwool are typical choices. Select a medium that encourages strong root growth and has good moisture retention.

3. Prepare the Reservoir: The nutrient solution that will be given to your plants is kept in the reservoir. Make sure the container is lightproof to stop algae growth and big enough to fit the amount of plants you want to grow in it. For irrigation by gravity, set the reservoir below the level of the plants.

4. Supply Mechanism: A method of supply that sends the nutrient-containing solution straight to the plants via tubing or pipelines may be necessary to guarantee that the plants are well-saturated and receiving the proper amount of nutrients. To prevent spillage, make sure they are all securely linked.

5. Mix Nutrient Solution: Prepare a nutrient solution according to the specific requirements of your chosen marijuana strain and growth stage. Use a quality hydroponic nutrient solution and follow the manufacturer's instructions for mixing ratios and pH levels. Regularly monitor and adjust the nutrient solution as needed to maintain optimal plant health.

6. Transplant Seedlings or Cuttings: It's time to move your marijuana seedlings or cuttings into the growing medium after your hydroponic system is configured and the nutrient solution is ready. Make sure the roots are sufficiently covered and supported when you gently insert them into the medium. Throughout this procedure, take care not to harm the fragile roots.

7. Monitor Environmental Conditions: Maintain proper environmental conditions within your grow space, including temperature, humidity, and light levels. Invest in quality grow lights to provide adequate illumination for your plants, and use fans or ventilation systems to ensure air circulation and prevent mold or mildew growth.

8. Regular Maintenance: Monitor your hydroponic system regularly to ensure everything is functioning correctly. Check nutrient levels, pH levels, and water temperature daily and make adjustments as needed. Clean the reservoir and delivery system periodically to prevent clogs or buildup of algae or debris.

By following these steps and staying vigilant with maintenance and monitoring, you can set up a successful hydroponic system for cultivating marijuana and enjoy the benefits of increased yield, faster growth rates, and precise control over growing conditions.

Types of Hydroponic Systems for Marijuana

There are several varieties of hydroponic systems available for producing marijuana, and each has its own set of benefits and considerations. . The following is a summary of the most common types:

1. Plants grown using Deep Water Culture (DWC) systems: With this method, plants float while their roots are immersed in a nutrient solution. The fluid is oxygenated with the use of bubblers, which promotes the growth of healthy roots. Though DWC is simple to set up and suitable for beginners, pH and nutrient levels must be regularly monitored..

2. Nutrient Film Technique (NFT): NFT systems utilize a shallow, sloped channel to continuously recirculate a thin film of nutrient solution over the plant roots. This method provides ample oxygen to the roots and is highly efficient in water and nutrient usage. However, it requires precise adjustments to prevent clogging and ensure uniform nutrient distribution.

3. Ebb and Flow (Flood and Drain): Ebb and Flow systems periodically flood the growing medium with nutrient solution before draining it away. This cycle ensures that roots receive both water and oxygen, promoting healthy growth. Ebb and Flow systems are versatile and suitable for various plant sizes but require careful timing to prevent over- or under-watering.

4. Drip Systems: Drip systems deliver nutrient solution directly to the plant's root zone through drip emitters or lines. This method allows for precise control over nutrient delivery and is suitable for larger plants or those with high water requirements. Drip systems are customizable and adaptable to different growing setups but may require more maintenance to prevent clogging.

5. Aeroponics: Aeroponic systems mist nutrient solution onto the plant roots suspended in the air. This method maximizes oxygen exposure to the roots, promoting rapid growth and nutrient uptake. Aeroponic systems are highly efficient in water and nutrient usage and are suitable for compact or vertical growing setups but require careful monitoring to prevent drying out or clogging of misters.

Regardless of the system chosen, proper setup, maintenance, and monitoring are crucial for successful hydroponic marijuana cultivation.

Choosing the Right System for Your Needs

Choosing the right hydroponic system for your needs requires careful consideration of several factors, including your experience level, available space, budget, and specific growing requirements. Here's a guide to help you select the most suitable system:

1. **Experience Level:** If you're new to hydroponic gardening, consider starting with a simpler system that's easier to set up and maintain, such as Deep Water Culture (DWC) or a basic Drip System. These systems are beginner-friendly and require minimal technical knowledge. As you gain experience and confidence, you can explore more advanced systems like Nutrient Film Technique (NFT) or Aeroponics.

2. **Space Availability:** Determine how much room you have for your hydroponic system. If you're short on room, think about compact or vertical systems like NFT or Aeroponics, which make effective use of vertical growth space. As an alternative, larger systems like Ebb and Flow or Drip Systems, which enable more extensive cultivation, can be chosen if you have enough room.

3. **Budget:** Ascertain how much money you'll need to install and run your hydroponic system. Certain systems, such as DWC or simple drip systems, are reasonably priced and need little equipment investment. On the other hand, because they require specialized parts and technology, more sophisticated systems like NFT or Aeroponics could have more upfront expenditures.

4. **Growing Requirements:** Consider the specific needs of the plants you intend to grow, such as their size, growth rate, and water/nutrient requirements. Certain systems may be better suited for specific types of plants or growth stages. For example, Aeroponic

systems are ideal for rapid-growing plants like lettuce, while DWC systems are well-suited for larger plants like tomatoes or cannabis.

5.Maintenance and Monitoring: Evaluate how much time and effort you're willing to dedicate to maintaining and monitoring your hydroponic system. Some systems, like DWC or Ebb and Flow, require regular monitoring of nutrient levels and pH, while others, like NFT or Aeroponics, may require more frequent maintenance to prevent clogging or drying out of misters.

6.Climate and Environmental Factors: Consider your local climate and environmental conditions when selecting a hydroponic system. Some systems, like Aeroponics, may be more sensitive to temperature and humidity fluctuations and require additional climate control measures. Others, like DWC or Ebb and Flow, may be more forgiving and adaptable to a wider range of conditions.

By carefully evaluating these factors and choosing the hydroponic system that best aligns with your needs and preferences, you can set yourself up for success in cultivating healthy and thriving plants. Don't hesitate to seek advice from experienced growers or consult with hydroponic experts to ensure you make an informed decision.

1.2 UNDERSTANDING NUTRIENTS

Essential Nutrients for Marijuana Growth

Like all other plants, marijuana plants need certain nutrients to sustain healthy growth and development over the course of their lifespans. Macronutrients and Micronutrients are the two groups into which these nutrients can be separated. The following is a summary of the vital nutrients needed for the growth of marijuana:

a.Nitrogen (N): For plants to produce leaves and be healthy overall, Nitrogen is essential. It is necessary for the synthesis of chlorophyll, which is necessary for photosynthesis. A lack of nitrogen can cause growth retardation and leaf yellowing.

b. Phosphorus (P): Root growth, flowering, and fruiting all depend on phosphorus. It is especially crucial during the flowering stage and helps the plant transport energy within itself. Reduced flower production and poor root growth might result from phosphorus deficit.

c. Potassium (K): In plants, potassium controls the intake of water, the movement of nutrients, and the activation of enzymes. It enhances disease resistance and fortifies cell walls. A lack of potassium can lead to weakened stems, poor flowering, and an increased vulnerability to pests and illnesses.

d. Calcium (Ca): Membrane integrity and the structure of cell walls depend on calcium. It enhances general plant health and aids in the prevention of diseases like bloom end rot. Leaf tissue necrosis and distorted growth can be caused by a calcium shortage.

e. Magnesium (Mg): Magnesium has a role in photosynthesis and enzyme activation. It is a component of chlorophyll. It also affects how nutrients are absorbed and how carbohydrates are metabolized. Older leaves that have chlorosis (yellowing) may be the result of magnesium shortage.

f. Sulfur (S): In plants, sulfur is necessary for the synthesis of proteins, vitamins, and amino acids. Enzyme activity and nutrient absorption are aided by it. Reduced growth and leaf yellowing might be symptoms of a sulfur deficit.

g. Iron (Fe): In photosynthesis, iron is necessary for the production of chlorophyll and the movement of electrons. Iron deficiency can lead to chlorosis in young leaves.

h. Manganese (Mn): Photosynthesis, the activation of enzymes, and nitrogen metabolism are all impacted by manganese. Reduced development and interveinal chlorosis can result from a manganese deficit.

i. Zinc (Zn): Protein synthesis, hormone synthesis, and enzyme activation all depend on zinc. Growth retardation and deformed leaf growth might result from zinc deficiency.

j. Copper (Cu): Copper is involved in enzyme activation and electron transport. Copper deficiency can cause wilting and necrosis of leaf tips.

k. Boron (B): Boron is essential for cell wall formation, hormone regulation, and reproductive development. Boron deficiency can result in brittle stems and distorted growth.

l. Molybdenum (Mo): Molybdenum is required for nitrogen fixation and enzyme activity. Molybdenum deficiency can cause yellowing and cupping of leaves.

Ensuring that marijuana plants receive a balanced nutrient solution with sufficient macronutrients and micronutrients is vital for optimizing growth, yield, and plant health. Careful monitoring of nutrient levels and adjusting the nutrient solution as necessary are critical to avoid deficiencies or excesses.

Nutrient Solutions and Mixtures

Nutrient solutions play a crucial role in hydroponic marijuana cultivation, providing essential elements for healthy plant growth. Let's explore the key aspects of hydroponic nutrients for cannabis:

Importance of Nutrients:

Proper nutrients are vital for healthy cannabis growth. Without them, plants can become weak or even die.

Nutrients are especially critical during different growth stages:

Seedling Stage: High levels of nitrogen, phosphorus, and potassium (NPK) are needed for root development and leaf growth.

Vegetative Stage: As plants grow taller, they require less nitrogen and more phosphorus and potassium for stem strength and bud formation.

Flowering Stage: During bud development, higher levels of potassium and phosphorus are essential.

The specific nutrient requirements can vary based on the cannabis strain and environmental factors.

Below is a brief overview of nutrients mixtures and solutions that are often employed in hydroponic marijuana cultivation:

1. Commercial Nutrient Formulations: They are a boon for hydroponic marijuana growers. These pre-formulated nutrient solutions, tailored for cannabis cultivation, come in liquid or powder form. They contain a balanced blend of essential nutrients:

Macronutrients:

Nitrogen: Vital for leaf growth and overall plant health.
Phosphorus: Crucial for root development and flowering.
Potassium: Supports overall plant vigor.

Micronutrients:

Iron, manganese, and zinc: Essential for various metabolic processes. These formulations are convenient and ensure that cannabis plants receive all necessary nutrients in the correct proportions.

2. Two-Part Nutrient Systems: Some growers prefer to use two-part nutrient systems, which consist of separate bottles of "Grow" and "Bloom" solutions. The Grow solution is used during the vegetative stage to promote lush foliage and root development, while the Bloom solution is used during the flowering stage to encourage flower formation and bud development. These systems allow for more precise control over nutrient ratios at different growth stages.

3. Custom Nutrient Mixtures: Experienced growers may choose to mix their own nutrient solutions using individual fertilizer salts or concentrates. This approach allows for greater customization and fine-tuning of nutrient levels to suit specific strains or growing conditions. Common fertilizer salts used in custom nutrient mixtures include calcium nitrate, potassium phosphate, magnesium sulfate, and various chelated micronutrients.

4. Organic Nutrient Solutions: Organic growers may opt for nutrient solutions made from natural and organic sources, such as compost teas, seaweed extracts, fish emulsions, or guano-based fertilizers. These organic nutrients provide a slower release of nutrients and promote soil microbial activity, contributing to overall soil health and plant vitality. However, organic nutrient solutions may require more frequent monitoring and adjustments to maintain nutrient balance.

To prevent over- or under-fertilization, it's critical to carefully follow nutrient recipes or manufacturer guidelines while making nutrient solutions. To guarantee the best possible nutrient uptake by the plants, you should also regularly monitor the pH and electrical conductivity (EC) of the fertilizer solution and make any necessary adjustments. In order to prevent nutritional deficiencies or toxicities and to promote healthy plant growth and development, it is imperative to regularly monitor pH and nutrient levels.

Calculating and Adjusting Nutrient Levels

Calculating and adjusting nutrient levels in hydroponic systems involves understanding the nutrient requirements of your plants, measuring the concentration of nutrients in your solution, and making appropriate adjustments to maintain optimal levels. Here's a step-by-step guide:

1. Determine Nutrient Requirements: Understand the nutrient requirements of your marijuana plants at different growth stages. Refer to nutrient guidelines or recommendations specific to cannabis cultivation, considering factors such as plant age, growth rate, and environmental conditions.

2. Test Nutrient Solution: To measure the pH and EC of your nutrient solution, use an electrical conductivity (EC) meter and a pH meter. While EC measures the concentration of dissolved ions, including nutrients, pH assesses the solution's acidity or alkalinity.

3. Adjust pH: Use pH up or pH down solutions to bring your nutrient solution's pH within the ideal range for cannabis cultivation, which is normally between 5.5 and 6.5. After gradually adding small amounts of pH adjusters to the nutrient solution and stirring constantly, check the pH to make sure it is within the appropriate range.

4. Interpret EC Readings: EC readings indicate the concentration of nutrients in the solution. Compare your EC readings to target ranges recommended for marijuana cultivation. Adjustments may be necessary if the EC is too high (indicating over-fertilization) or too low (indicating under-fertilization).

5. Adjust Nutrient Levels: Depending on your EC readings and the nutrient requirements of your plants, adjust the nutrient levels in your solution accordingly. Increase nutrient concentration by adding more fertilizer or nutrient solution, or decrease it by diluting the solution with fresh water. Follow manufacturer recommendations or nutrient recipes to ensure accurate adjustments

6. Monitor and recheck: To ensure that levels are within the intended range, recheck the pH and EC of your nutrition solution after making any necessary modifications. Maintain optimal conditions by routinely monitoring nutrient levels and making necessary adjustments, particularly during times of rapid plant growth or environmental changes.

7. Maintain Records: Maintain thorough records of nutrient concentrations, pH changes, and plant reactions over time. You can use this data to monitor the success of your nutrient management strategies and plan forward by making well-informed decisions. Through adherence to these guidelines and consistent observation and modification of nutrients concentrations in your hydroponic setup, you can guarantee that your cannabis plants obtain the vital nutrients required for robust development and optimal productivity.

CHAPTER 2

SELECTING THE RIGHT STRAIN

2:1 INTRODUCTION TO MARIJUANA STRAINS

Cannabis, the scientific name for Marijuana, is a multipurpose plant that comes in a variety of strains, each having its own distinct effects, smells, fragrances, and therapeutic qualities. Both recreational marijuana users and medical patients looking for particular therapeutic benefits need to understand marijuana strains. An outline of the basic principles of cannabis strains is provided below:

1. Genetics: The classification of marijuana strains is done on the basis of their genetic heritage, which affects their overall effects, cannabinoid profiles, and growth characteristics. Indica, Sativa, and hybrid strains are the three main categories of cannabis varieties.

Indica: Indica strains are great for use in the evenings or for unwinding at night because of their well-known sedative and calming qualities. They usually have larger leaves, shorter stature, and thick buds. When it comes to reducing stress, anxiety, sleeplessness, and chronic pain, indicas are frequently chosen.

Sativa: Due to their elevating and invigorating qualities, sativa strains are good for use during the day or for creative endeavors. They often have narrower leaves, longer flowering seasons, and higher plants. It's normal practice to utilize sativas to improve focus, productivity, mood, and creativity.

Hybrid: Strains that combine the traits and effects of both parent strains—Indica and Sativa genetics—are called hybrids. The effects of hybrids can differ greatly; some are more Indica-dominant (hybrids with a strong Indica influence), while others are more Sativa-dominant (hybrids with a strong Sativa influence).

2. Cannabinoid Profile: Cannabis strains are known to possess a diverse range of cannabinoids, including THC (tetrahydrocannabinol), CBD (cannabidiol), and other compounds, each with unique impacts and medical qualities. A strain's overall potency and potential for medicinal use are determined by the ratio of THC to CBD and other cannabinoids.

3. Terpene Profile: Cannabis contains aromatic molecules called terpenes, which add to the plant's flavor, scent, and effects. Every strain has a different terpene profile, which can affect both the way it feels and works as a medicine. Myrcene, limonene, pinene, and linalool are notable terpenes present in cannabis.

4. Uses and Effects: Depending on one's needs and tastes, different cannabis strains can have a variety of effects. While many strains are prized for their psychoactive and intoxicating effects, others are appreciated for their therapeutic qualities, which include reducing anxiety, discomfort, inflammation, and stimulating appetite.

Growers should take into account many elements such as climate, indoor or outdoor growing conditions, space constraints, and cultivation complexity when choosing a strain of marijuana. Particular environments suit some strains better than others in terms of resilience and adaptability.

In summary, marijuana strains encompass a wide spectrum of genetics, effects, and characteristics, offering something for everyone, whether seeking recreational enjoyment, therapeutic relief, or medicinal benefits. Understanding the nuances of different strains empowers users and patients to make informed choices and find the strains that best suit their preferences and needs.

Indica vs. Sativa vs. Hybrid

When selecting Indica, Sativa, or Hybrid strains for hydroponic cultivation, various factors must be taken into account to achieve optimal outcomes. The following is a comparison of Indica, Sativa, and Hybrid strains within the realm of hydroponic growing:

1. Indica:

Suitability: Indica strains are typically ideal for hydroponic cultivation because of their compact stature and brief flowering cycles. Their dense leaves and robust stems make them adaptable to a range of hydroponic setups.

Growth Characteristics: Indica plants typically exhibit a bushy, compact growth pattern with broader leaves and shorter internodal spacing. This growth habit makes them suitable for training techniques like topping, pruning, and low-stress training (LST) to maximize yields in limited vertical space.

Flowering Time: Growers aiming to reduce crop cycles and get faster harvests in hydroponic systems may find that Indica strains have shorter flowering times than Sativas.

Effects: Indica strains are popular for use at night or as medicine because of its soothing, calming properties. They might be useful in reducing chronic pain, stress, anxiety, and insomnia.

2. Sativa:

Suitability: Sativa strains can also be cultivated hydroponically, but their taller stature and longer flowering periods may require additional space and support structures. Sativas may benefit from vertical hydroponic setups or taller grow spaces.

Growth Characteristics: Sativa plants typically exhibit tall, lanky growth with longer internodal spacing and narrower leaves. Growers may need to manage height and stretch during the vegetative stage to prevent overcrowding and light penetration issues in hydroponic systems.

Flowering Time: Sativa strains often have longer flowering times compared to Indicas, requiring patience and careful monitoring throughout the growth cycle. However, some Sativa-dominant hybrids may offer a compromise with shorter flowering periods.

Effects: Sativa strains are associated with uplifting, cerebral effects that promote creativity, energy, and focus. They are favored for daytime use or stimulating activities and may help alleviate mood disorders, fatigue, and lack of appetite.

3.Hybrid:

Suitability: Hybrid strains, which combine traits of both Indica and Sativa genetics, can be cultivated in hydroponic systems with careful consideration of their individual characteristics. Depending on the specific hybrid, growers may need to adjust cultivation techniques and environmental conditions accordingly.

Growth Characteristics: Hybrid plants may exhibit a combination of growth traits from their parent strains, offering a balance between Indica's compact structure and Sativa's taller stature. Growers can choose hybrids based on their desired growth habits and cultivation preferences.

Flowering Time: Flowering times of hybrid strains can vary widely depending on their genetic makeup, with some leaning towards Indica-dominance or Sativa-dominance. Select hybrids with flowering times that fit your schedule and production goals.

Effects: Hybrids offer a diverse range of effects, blending the relaxing, therapeutic properties of Indicas with the uplifting, cerebral effects of Sativas. Choose hybrids based on their cannabinoid and terpene profiles to target specific effects or medicinal benefits.

In conclusion, all three types of marijuana strains—Indica, Sativa, and Hybrid—can be successfully cultivated in hydroponic systems with proper care, attention to detail, and cultivation techniques tailored to their specific characteristics. Growers should consider factors such as growth habits, flowering times, effects, and suitability for their growing environment when selecting strains for hydroponic cultivation.

Understanding Cannabinoid Content

To fully comprehend the cannabinoid content of cannabis grown hydroponically, one must have a working knowledge of the ways in which genetics, environment, and growing techniques influence the development of cannabinoids such as THC (tetrahydrocannabinol) and CBD (cannabidiol). The following is a summary of the main things to think about:

1. Genetics:

Strain Selection: The production of cannabinoids is significantly influenced by the genetic makeup of a chosen marijuana strain. Different strains have varying THC, CBD, and other cannabinoid concentrations. Selecting a strain for hydroponic development requires careful consideration of the strain's intended effects and medical benefits in addition to its cannabinoid profile

2. Environmental Conditions:

Light: The duration, spectrum, and intensity of light are important factors in the synthesis of cannabinoids. Cannabinoid production can be increased by providing ideal lighting conditions that are customized to meet the strain's unique requirements. In hydroponic systems, LED grow lights with programmable spectra are frequently utilized to control light exposure.

Temperature: Throughout the growing cycle, it is critical to maintain consistent temperatures that fall within the ideal range, which is normally between 70-85°F or 21-29°C. This will maximize the production of cannabinoids. Temperature variations can stress plants and have an impact on the synthesis of cannabinoids

Humidity: Humidity plays a crucial role in cannabis cultivation, affecting both plant health and cannabinoid production. 40-60% humidity is considered optimal for cannabis growth. During the flowering stage, excessive humidity can lead to bud rot, which can ruin the crop. Insufficient humidity may cause stress to the plants and reduce resin production.

Maintaining the right humidity levels is essential for successful cannabis cultivation and maximizing cannabinoid yield.

Air Quality: Good air circulation and ventilation are essential for maintaining healthy plant growth and optimizing cannabinoid production. CO_2 supplementation may also enhance cannabinoid synthesis during the flowering stage when plants have increased metabolic demands.

3. Nutrient Management:

Nutrient Solution: Balancing nutrient levels and maintaining optimal pH in the hydroponic system is critical for healthy plant growth and cannabinoid production. Providing a well-rounded nutrient solution with essential macronutrients (N-P-K) and micronutrients supports overall plant health and vigor.

Supplements: Some growers use specific nutrient supplements or additives purported to enhance cannabinoid production. These may include organic amendments, bloom boosters, or microbial inoculants. However, caution should be exercised, and products should be used judiciously to avoid nutrient imbalances or toxicity.

4. Cultivation Techniques:

Training: Pruning, topping, and other training techniques can influence plant structure and cannabinoid distribution. Techniques like low-stress training (LST) or ScrOG (screen of green) may improve light penetration and encourage more uniform cannabinoid production.

Harvest Timing: Harvesting at the optimal time when cannabinoid levels are highest maximizes potency and therapeutic benefits. Trichome maturity, bud density, and terpene profiles are indicators used to determine the ideal harvest window.

5. Monitoring and Testing:

Regular monitoring of plant health, nutrient levels, and environmental conditions is essential for identifying any issues that may impact cannabinoid production. Additionally, testing cannabinoid levels through laboratory analysis provides valuable insights into the potency and composition of the final product.

By understanding how these factors influence cannabinoid content, hydroponic growers can implement strategies to optimize cultivation practices and produce high-quality marijuana with desired cannabinoid profiles.

Factors to consider when choosing a strain

When selecting a strain for hydroponic cultivation, several factors should be taken into consideration to ensure successful growth and optimal yields. Here are some key factors to consider:

1. Genetics and Characteristics:

Indica, Sativa, or Hybrid: Decide whether you want to grow an Indica, Sativa, or Hybrid strain based on your preferences and desired effects. Consider factors such as plant height, leaf structure, and growth patterns associated with each type.

Cannabinoid Profile: Choose a cannabis strain whose profile of cannabinoids aligns with the intended use. Whether you're searching for balanced THC/CBD ratios for medicinal purposes or high THC levels for recreational use, pick the strain that best meets your needs

Terpene Profile: Consider the terpene makeup of the strain, as terpenes affect the flavor, aroma, and effects of cannabis. Different terpenes may offer unique benefits and experiences.

Yield and Flowering Time: Look for information on the expected yield and flowering time of the strain. Select a strain that fits your timeline and production goals, whether you're aiming for a quick turnaround or maximum yields

Disease Resistance: Choose strains known for their resistance to common pests, diseases, and environmental stressors. Robust genetics can help minimize the risk of crop loss and ensure a successful harvest..

2. Growing Environment:

Indoor vs. Outdoor: Consider whether you'll be growing indoors or outdoors and choose a strain that's well-suited to your chosen environment. Some strains thrive in controlled indoor environments, while others are better suited for outdoor cultivation.

Space Constraints: Assess your available growing space and select a strain that fits within your space limitations. Choose compact, bushy strains for smaller grow spaces and taller, more expansive strains for larger areas.

3.Cultivation Difficulty:

Experience Level: Consider your level of experience with hydroponic cultivation. If you're a beginner, choose a strain known for its ease of cultivation and forgiving nature. Experienced growers may opt for more challenging strains with unique characteristics or higher potency.

4.Climatic Conditions:

Temperature and Humidity: Consider the conditions of your growing environment with regard to temperature and humidity. Select strains that can flourish in your climate and are compatible with the circumstances you may offer.

5.Individual Choices:

Impacts: Think about the impacts that a strain should have. Choose a strain that will give you the effects you want, whether they are calming and tranquil, stimulating and uplifting, or a combination of both.

Aroma and Flavor: Pick cannabis strains whose flavors and scents satiate your senses. Choose a strain with terpene profiles that suit your tastes, whether you like scents that are fruity, citrusy, earthy, or floral.

You can select a strain that meets your requirements and tastes while being ideal for hydroponic cultivation by carefully weighing these variables. Investigate thoroughly, speak with seasoned cultivators, and try out various strains to determine which is best for your hydroponic growing operation..

2.2 BEST STRAINS FOR HYDROPONIC CULTIVATION

Selecting the best strains for hydroponic cultivation involves considering factors such as growth characteristics, cannabinoid profiles, yield potential, and suitability for the hydroponic environment. While individual preferences and cultivation goals may vary, here are some strains that are generally well-suited for hydroponic cultivation:

1.Northern Lights:
Type: Indica
Characteristics: Northern Lights is a compact, resilient strain with a short flowering time, making it ideal for hydroponic setups with limited space. It produces dense, resinous buds with a sweet, earthy aroma.
Yield: Moderate to high yield
Effects: Known for its relaxing, sedative effects, Northern Lights is popular among medical cannabis patients seeking relief from insomnia, pain, and stress.

2.White Widow
Type: Hybrid (Indica-dominant)
Characteristics: White Widow is a versatile strain that performs well in hydroponic systems. It features a robust structure, making it suitable for various training techniques. It produces large, sticky buds with a pungent, spicy aroma.
Yield: High yield
Effects: White Widow offers a balanced combination of relaxation and mental stimulation, making it suitable for both recreational and medicinal users.

3. Blue Dream:
Type: Hybrid (Sativa-dominant)
Characteristics: Blue Dream is a vigorous, fast-growing strain that responds well to hydroponic cultivation. It has a tall, branchy structure and produces long, dense colas with a sweet, fruity aroma.

Yield: High yield

Effects: Blue Dream provides uplifting, euphoric effects coupled with gentle relaxation, making it a popular choice for daytime use and managing symptoms of depression, anxiety, and pain.

2. OG Kush:

Type: Hybrid (Indica-dominant)

Characteristics: OG Kush is a classic strain known for its potency and resin production, making it well-suited for hydroponic cultivation. It has a compact, bushy structure and produces dense, frosty buds with a distinctive earthy, piney aroma.

Yield: Moderate to high yield

Effects: OG Kush delivers a powerful, euphoric high accompanied by deep relaxation, making it suitable for evening use and managing symptoms of chronic pain, insomnia, and stress.

3. Jack Herer

Type: Sativa

Characteristics: Jack Herer is a popular Sativa strain known for its energetic growth and high resin production. It has a tall, slender structure and produces large, airy buds with a spicy, herbal aroma.

Yield: Moderate to high yield

Effects: Jack Herer offers an uplifting, creative high with clear-headed effects, making it suitable for daytime use and enhancing mood, focus, and productivity.

4. Girl Scout Cookies (GSC):

Type: Hybrid (Indica-dominant)

Characteristics: GSC is a high-yielding strain with a compact, bushy structure that adapts well to hydroponic cultivation. It produces dense, colorful buds with a sweet, earthy aroma.

Yield: High yield

Effects: GSC delivers a potent, euphoric high coupled with deep relaxation, making it suitable for both recreational and medicinal users seeking relief from pain, anxiety, and appetite loss.

These strains are renowned for their performance in hydroponic systems and have earned acclaim among growers for their growth characteristics, cannabinoid profiles, and overall quality. Experimenting with different strains and cultivation techniques can help you find the best options for your hydroponic grow operation.

High Yield Strains

In cannabis cultivation, choosing strains with high yields is essential. Such strains often produce more flowers, leading to greater total yields. Below are some well-known high-yield strains recognized for their substantial production:

1. Critical Mass:
Type: Indica-dominant hybrid
Characteristics: Critical Mass is renowned for its substantial yields and vigorous, rapid growth. It yields dense, resin-coated buds that exude a sweet, earthy fragrance. This variety is fairly simple to cultivate and thrives in a range of growing conditions.
Yield: Exceptionally high yield.

2. Big Bud:
Type: Indica-dominant hybrid
Characteristics: As the name suggests, Big Bud is renowned for its exceptionally large, heavy buds. This strain is known for its vigorous growth and high resin production, resulting in impressive yields. Big Bud has a sweet, spicy aroma and offers relaxing effects.
Yield: Very high yield

3. Super Silver Haze:
Type: Sativa-dominant hybrid
Characteristics: Super Silver Haze is prized for its towering colas and abundant resin production. Despite its Sativa dominance, this strain can still deliver significant yields, especially when grown under optimal conditions. Super Silver Haze has a complex flavor profile with hints of citrus, spice, and earth.
Yield: High yield

4. Amnesia Haze:
Type: Sativa-dominant hybrid

Characteristics: This strain is renowned for its energizing effects and sky-high yields; it yields long, dense colas packed with resin, making harvests impressive; it smells pungent and earthy with citrus and spice notes..

Yield: High yields

5. White Widow:

Type: Hybrid, mostly Indica

Characteristics: White Widow is a traditional strain that is well-liked for its strong effects and large yields. It yields big, resinous blooms with a strong, earthy scent and a robust structure. Growing White Widow is comparatively simple, and it responds well to many growth methods.

Yield: High yield.

6. Gorilla Glue #4:

Type: Hybrid, with a dominant Sativa

Characteristics: Gorilla Glue #4 is a high-yielding strain with strong effects and dense, sticky buds..It produces abundant resin, making it ideal for concentrate production The flavor profile of Gorilla Glue #4 is complex, containing notes of citrus, diesel, and pine.

7. Green Crack:

Type: Sativa-dominant hybrid

Characteristics: Green Crack is prized for its vigorous growth and abundant yields. It produces long, resinous colas with a sweet, fruity aroma. Green Crack offers uplifting, energizing effects that are perfect for daytime use.

Yield: High yield

When cultivating high-yield strains, it's essential to provide optimal growing conditions, including adequate light, nutrients, water, and ventilation. Additionally, training techniques such as topping, pruning, and trellising can help maximize yields by promoting even canopy development and light penetration. With proper care and attention, these high-yield strains have the potential to deliver bountiful harvests for growers.

Disease-Resistant Strains

Disease-resistant strains are varieties of cannabis bred specifically to withstand common pests, pathogens, and environmental stressors, reducing the likelihood of crop loss and maintaining plant health. These strains possess genetic traits that enhance their resilience to various diseases and ailments, ensuring a higher likelihood of successful cultivation. Here are a few examples of disease-resistant strains:

1. CBD Critical Mass:
Type: Indica-dominant hybrid
Characteristics: CBD Critical Mass is renowned for its high CBD levels and remarkable resistance to disease. It yields dense, resin-coated buds that have an even THC to CBD ratio, making it ideal for medical users seeking alleviation from pain, inflammation, and anxiety.
Disease Resistance: CBD Critical Mass demonstrates resistance to common pests and diseases, offering growers a dependable option for a low-maintenance strain with a reduced risk of crop failure.

2. Dance World:
Type: Sativa-dominant hybrid
Characteristics: Dance World is highly valued for its uplifting, euphoric effects and high CBD concentration. It yields big, resinous buds that have a subtle citrus flavor and a pleasant, fruity scent. Dancing World works effectively for reducing fatigue, tension, and depressive symptoms during the day.
Disease Resistance: Dance World exhibits strong disease resistance, rendering it a sturdy choice for hydroponic farming. It is capable of resisting typical pests and diseases, ensuring vigorous growth during the entire cultivation cycle.

3. Super Silver Haze:
Type: Sativa-dominant hybrid

Characteristics: Super Silver Haze is celebrated for its potent effects and abundant resin production. It produces long, dense colas with a spicy, citrusy aroma and a complex flavor profile. Super Silver Haze offers uplifting, energizing effects that are perfect for creative endeavors and social gatherings.

Disease Resistance: Super Silver Haze exhibits strong disease resistance, making it a reliable choice for hydroponic growers. It can withstand environmental stressors and maintain healthy growth, resulting in impressive yields and high-quality buds.

4.Blue Dream:

Type: Hybrid (Sativa-dominant)

Characteristics: Blue Dream is renowned for its fast growth and abundant yields. It produces long, dense colas with a sweet, berry-like aroma and a smooth, fruity flavor. Blue Dream offers balanced effects, combining relaxation with gentle euphoria, making it suitable for day or night use.

Disease Resistance: Blue Dream is known for its resilience to common pests and diseases, making it a popular choice for hydroponic cultivation. It can withstand adverse conditions and maintain vigorous growth, resulting in healthy, high-yielding plants. These disease-resistant strains are excellent options for hydroponic cultivation, offering growers peace of mind and ensuring successful harvests with minimal risk of crop loss. By incorporating these resilient varieties into their cultivation operations, growers can maintain plant health, increase yields, and produce high-quality cannabis consistently.

Fast growing strains

Fast-growing strains are well-suited for hydroponic marijuana cultivation, enabling growers to harvest more quickly and increase efficiency. Below is a summary of some popular fast-growing strains that are appropriate for hydroponic systems:

1.Northern Lights:

Type: Indica
Characteristics: The Northern Lights strain is renowned for its swift growth and brief flowering period. It yields dense buds rich in resin, exuding a sweet and earthy scent. This variety is fairly simple to cultivate and is well-suited for hydroponic setups.
Yield: Moderate to high yield

2. White Widow:
Type: Hybrid (Indica-dominant)
Characteristics: White Widow is a versatile strain that performs well in hydroponic setups. It features a robust structure and produces large, sticky buds with a pungent, spicy aroma. White Widow has a relatively short flowering time, making it suitable for quick turnaround.
Yield: High yield

3. Blue Dream:
Type: Hybrid (Sativa-dominant)
Characteristics: Blue Dream is celebrated for its vigorous growth and fast flowering. It has a tall, branchy structure and produces long, dense colas with a sweet, fruity aroma. Blue Dream adapts well to hydroponic systems and offers high yields.
Yield: High yield

4. Amnesia Haze:
Type: Sativa-dominant hybrid
Characteristics: Amnesia Haze is prized for its towering colas and rapid growth. It produces long, dense buds with a pungent, earthy aroma. Despite its Sativa dominance, Amnesia Haze has a relatively short flowering time, making it ideal for hydroponic cultivation.
Yield: High yield

5. Green Crack:
Type: Sativa-dominant hybrid
Characteristics: Green Crack is known for its vigorous growth and abundant yields. It produces long, resinous colas with a sweet, fruity aroma. Green Crack offers a relatively short flowering time and delivers uplifting, energizing effects.

Yield: High yield

These fast-growing strains are well-suited for hydroponic cultivation due to their rapid growth rates, short flowering times, and high yields. By selecting these strains and providing optimal growing conditions, growers can achieve quick and successful harvests in their hydroponic setups.

CHAPTER 3

ENVIRONMENT AND CLIMATE CONTROL

3.1 OPTIMIZING GROWING CONDITIONS

Temperature and Humidity Control

Temperature and humidity control are essential aspects of hydroponic marijuana cultivation, as they directly influence plant health, growth, and overall yield. Maintaining optimal environmental conditions ensures that plants can thrive and reach their full potential. Here's a summary of temperature and humidity control in hydroponic marijuana cultivation:

1. Temperature Control:

Optimal Range: The ideal temperature range for hydroponic marijuana cultivation typically falls between 70-85°F (21-29°C) during the day and slightly cooler at night. Maintaining stable temperatures within this range promotes healthy growth and metabolic processes.

Climate Control Systems: Various climate control systems, such as heaters, air conditioners, and ventilation systems, help regulate temperatures in indoor grow environments. Automated systems with temperature sensors and controllers allow growers to monitor and adjust conditions as needed.

Heat Management: Heat management is crucial, especially in confined grow spaces where temperatures can rise rapidly. Adequate airflow, ventilation, and strategic placement of fans help dissipate heat and maintain optimal conditions. Additionally, using heat-resistant materials for grow room construction can minimize heat buildup.

Seasonal Adjustments: Growers may need to make seasonal adjustments to temperature control systems to accommodate changes in ambient temperatures. Summer heat waves or winter cold spells may require additional cooling or heating measures to keep conditions within the optimal range.

2. Humidity control:

Optimal Range: Depending on the stage of growth, there are different humidity ranges that work best for hydroponic cannabis growing.

Vegetative stage: Relative humidity should be kept between 40 and 70 percent throughout the vegetative stage. This range encourages plant transpiration and growth.

Flowering Stage: To avoid problems like mold and bud rot during blooming, RH levels should be lower, ideally between 40 and 50%

Humidifiers and Dehumidifiers: Humidifiers and dehumidifiers are essential tools for controlling humidity levels in indoor grow environments. Humidifiers increase moisture levels when RH is too low, while dehumidifiers remove excess moisture when RH is too high..

Air Circulation: Adequate air circulation and ventilation are crucial for maintaining consistent humidity levels across the growing area. The use of oscillating and exhaust fans enhances airflow, which helps to prevent the formation of stagnant air pockets and layers of varying humidity.

Condensation Prevention: In high-humidity environments, condensation can lead to mold growth and other complications. Insulating walls, floors, and ceilings, along with utilizing waterproof materials, can help prevent the buildup of condensation.

Efficient control of temperature and humidity in hydroponic environments is crucial for setting optimal conditions for plant growth. By consistently monitoring, employing appropriate equipment, and executing timely modifications, cultivators can secure a vigorous and bountiful harvest during the entire growth period.

Light Requirements

Light is one of the most critical factors in hydroponic marijuana cultivation, as it drives photosynthesis, plant growth, and ultimately, yield. Understanding the light requirements of cannabis plants is essential for achieving optimal results. Here's a summary of light requirements in hydroponic marijuana cultivation:

1.Intensity:
Optimal growth and yield in cannabis plants require intense lighting. Light intensity is commonly measured by photosynthetic photon flux density (PPFD), in micromoles per square meter per second ($\mu mol/m2/s$). During the vegetative stage, cannabis plants thrive under light levels between 400 to 600 $\mu mol/m^2/s$. For the flowering stage, a higher light intensity of 600–1000 $\mu mol/m^2/s$ promotes bud development. Adequate light intensity is vital for the plants to absorb the energy necessary for photosynthesis, resulting in rapid growth and larger yields

2.Spectrum:
Cannabis plants need a complete spectrum of light, encompassing blue, green, and red wavelengths, to facilitate different growth phases. Blue light (400-500 nm) encourages vegetative growth and robust foliage, while red light (600-700 nm) is crucial for flowering and bud production. LED grow lights are favored by hydroponic growers due to their customizable spectra, which can be adjusted for various growth stages. These full-spectrum LEDs optimize photosynthesis and crop yield by meeting the specific light

requirements of the plants. Additionally, some growers complement natural sunlight with artificial lights to maintain consistent lighting and ideal conditions year-round, particularly in indoor growing setups.

3. Duration:
Cannabis plants need a consistent light cycle for vegetative or flowering growth. In the vegetative phase, 18-24 hours of light daily (18/6 or 24/0) is typical to encourage strong vegetative growth and avert early flowering. For the flowering phase, a 12-hour light and 12-hour dark cycle (12/12) is necessary to start flowering and enhance bud formation. A strict adherence to this light schedule is vital to ensure successful flowering and optimal yield.

4. Uniformity:
Ensuring uniform light distribution throughout the canopy is essential for maximizing photosynthetic efficiency and preventing uneven growth. Proper spacing of light fixtures, reflective surfaces, and strategic placement of plants help achieve uniform light distribution and minimize shading.

Using techniques such as low-stress training (LST), topping, and pruning can also help promote an even canopy and optimize light penetration, resulting in more consistent growth and higher yields. By meeting the light requirements of cannabis plants through proper intensity, spectrum, duration, and uniformity, hydroponic growers can optimize plant growth, enhance productivity, and achieve high-quality yields. Regular monitoring of light conditions and adjustments as needed are key to maximizing the potential of the crop throughout the cultivation cycle.

CO2 Levels

Carbon dioxide (CO_2) is vital in hydroponic marijuana cultivation, serving a key role in photosynthesis—the process through which plants transform light into chemical energy. Grasping and regulating the ideal CO_2 levels are critical for enhancing plant growth, yield, and productivity. Below is a summary of CO2 levels in hydroponic marijuana cultivation:

1. Optimal Levels:

CO2 The optimal levels of concentration for growing cannabis usually falls between 1000 to 1500 parts per million (ppm) throughout the lighting period. Such increased CO2 levels boost the rate of photosynthesis, resulting in quicker growth and larger harvests. During the dark phase, it's advisable to let CO2 concentrations diminish naturally since plants cease photosynthesis without light, often returning to normal ambient levels around 400 ppm.

2. Supplementation:

In indoor grow environments, CO2 supplementation is often necessary to maintain optimal levels, especially in enclosed spaces with limited ventilation. CO2 supplementation methods include:

CO2 generators or burners: Propane or natural gas burners that produce CO2 as a byproduct of combustion.

Compressed CO2 tanks: Pressurized tanks containing CO2 gas, which can be released into the grow environment using regulators and valves.

CO2 emitters or bags: Devices or bags containing CO2-releasing substances that gradually release CO2 over time.

3. Monitoring and Control:

Consistent monitoring of CO2 levels is crucial to ensure that plants get enough CO2 for their best growth. Portable CO2 monitors or controllers can help measure and regulate CO2 levels within the preferred range. Growers must also take into account other factors like temperature, humidity, and light intensity, since these can influence CO2 absorption and the rate of photosynthesis. Keeping these environmental conditions in equilibrium is key to fully leveraging the advantages of CO2 enrichment.

4. Safety Considerations:

When employing CO2 supplementation methods, it is crucial to follow safety precautions to mitigate potential risks such as CO2 accumulation and oxygen depletion. Ensuring adequate ventilation and correct installation of CO_2 equipment is vital for maintaining safe indoor air quality. Growers must also recognize the exposure limits of CO_2 for humans and maintain CO_2 concentrations within safe parameters to prevent health hazards.

Effective understanding and control of CO_2 levels in hydroponic marijuana cultivation allow growers to enhance photosynthesis, encourage robust growth, and increase yields. Consistent monitoring, appropriate supplementation, and strict adherence to safety protocols are essential for leveraging the advantages of CO_2 in promoting plant health and productivity.

3.2 MANAGING AIR CIRCULATION

Importance Of Airflow

Airflow is a crucial aspect of hydroponic marijuana cultivation, as it plays several vital roles in maintaining optimal growing conditions and promoting healthy plant growth. Here's a summary of the importance of airflow in hydroponic marijuana cultivation:

1. Temperature Regulation:
Proper airflow helps regulate temperature levels within the grow space by dissipating excess heat generated by grow lights, electronic equipment, and plant respiration. Adequate ventilation prevents heat buildup, reducing the risk of heat stress and maintaining optimal growing conditions for cannabis plants.

2. Humidity Control:
Airflow promotes humidity control by preventing moisture buildup and maintaining uniform humidity levels throughout the grow space. Proper ventilation helps reduce humidity in areas prone to excess moisture, preventing issues such as mold, mildew, and bud rot. Additionally, airflow aids in preventing condensation on plant surfaces, which can lead to fungal diseases and other problems.

3. CO2 Distribution:
Efficient airflow ensures the uniform distribution of carbon dioxide (CO_2) within the grow space, allowing plants to access sufficient CO_2 for photosynthesis. Proper CO_2 distribution enhances plant growth and productivity, leading to increased yields and improved quality of cannabis buds.

4. Oxygen Exchange:
Adequate airflow facilitates oxygen exchange, ensuring that plants receive an ample supply of oxygen for respiration. Oxygen is essential for root health and nutrient uptake, as well as for supporting beneficial microorganisms in the growing medium. Proper oxygen exchange promotes healthy root development and prevents issues such as root rot and anaerobic conditions.

5. Strengthening Stems:
Gentle airflow helps strengthen plant stems by providing a slight resistance that stimulates plant growth. As plants sway in the breeze, they develop thicker, sturdier stems, which can support heavier buds and improve overall plant structure.

6. Pest and Disease Prevention:
Proper airflow helps deter pests such as fungus gnats, spider mites, and aphids by creating a less hospitable environment for them to thrive. Additionally, airflow reduces the likelihood of fungal diseases by preventing stagnant air and excess moisture, which are conducive to fungal growth

7. Pollination:
In the case of outdoor or greenhouse cultivation, airflow aids in pollination by facilitating the movement of pollen between male and female cannabis plants. Adequate airflow ensures that pollen is distributed evenly, promoting successful pollination and seed development.

Overall, airflow is essential for creating a healthy and productive growing environment in hydroponic marijuana cultivation. By ensuring proper ventilation and air circulation, growers can optimize growing conditions, prevent common problems, and maximize the success of their cannabis crops. Regular monitoring of airflow, along with adjustments as needed, is key to maintaining optimal conditions throughout the cultivation cycle.

Preventing Molds and Mildew

In the cultivation of hydroponic marijuana, it is crucial to prevent mold and mildew to safeguard plant health, prevent crop loss, and guarantee high-quality yields. Mold and mildew thrive in environments with too much moisture, insufficient airflow, and inadequate light. The following is a summary of methods to prevent mold and mildew in hydroponic marijuana cultivation:

1. Optimize Air Circulation:
To ensure adequate airflow in the grow space, strategically position oscillating and exhaust fans. Adequate ventilation is essential to maintain steady air circulation, preventing the development of stagnant air pockets that could foster mold and mildew. Position the fans to promote airflow around the plants and minimize humidity buildup in the canopy areas.

2. Control Humidity Levels:
It's crucial to monitor and regulate humidity levels in the grow area to prevent mold and mildew. During the vegetative phase, maintain humidity under 60%, and aim for 40-50% during the flowering phase to reduce the likelihood of fungal infections. Employ dehumidifiers to extract surplus moisture and sustain ideal humidity conditions.

3. Ensure Proper Drainage:
In hydroponic systems, it's crucial to implement effective drainage to prevent water buildup and runoff. Make sure that containers or trays used for growing have sufficient drainage openings, allowing the surplus nutrient solution to flow out easily. It's also important to prevent overwatering, as overly moist growing media can foster the development of mold and mildew.

4. Maintain Cleanliness:

Keep the grow space clean and free of debris to minimize potential sources of mold and mildew. Regularly sanitize equipment, surfaces, and grow containers to remove organic matter and prevent microbial growth. Prune and remove dead or decaying plant material promptly to reduce the risk of fungal infections spreading.

5. Utilizing Beneficial Microorganisms:
Introducing beneficial microorganisms such as Bacillus subtilis and Trichoderma spp. into the growth medium can help suppress fungal pathogens and improve plant health. These advantageous microbes are capable of outcompeting harmful fungi for nutrients, colonizing the root zones, and stimulating the plant's immune response, thus reducing the risk of mold and mildew outbreaks.

6. Monitoring environmental conditions
Regularly tracking factors such as temperature, humidity, and CO_2 levels is a proactive approach that helps in identifying and resolving potential problems. Employing tools like hygrometers and thermometers is key to staying informed about these conditions and making necessary adjustments to preserve an ideal growth environment.

7. Regular plant inspection:
This is essential to identify signs of mold, mildew, or other fungal diseases. Be on the lookout for symptoms like white powdery substances, gray mold, or dark spots on leaves and stems. It's important to remove and dispose of any affected plant parts immediately to stop the spread of disease to healthy plants.

Hydroponic growers can reduce the incidence of mold and mildew by adopting these preventive practices, thus maintaining a healthy growing environment and promoting the well-being of their cannabis plants. Diligent monitoring, ensuring proper airflow, and maintaining strict cleanliness are crucial in preventing fungal infections and safeguarding plant health during the entire growth cycle.

Ventilation Systems

Ventilation systems are essential components of hydroponic marijuana cultivation setups, playing a critical role in maintaining optimal growing conditions, promoting plant health, and maximizing yields. Here's a summary of ventilation systems used in hydroponic marijuana cultivation:

1. Exhaust Fans:

Exhaust fans are key components of hydroponic grow rooms and grow tents, responsible for removing stale air, excess heat, and humidity from the environment. These fans draw air out of the grow space, creating negative pressure and promoting airflow.

Properly sized exhaust fans should be selected based on the size of the grow space and the amount of heat and moisture generated by grow lights, electronic equipment, and plant transpiration.

Exhaust fans are typically installed near the top of the grow space to remove warm air and humidity, venting it outside the building or through a designated exhaust system.

2. Intake Fans:

Intake fans complement exhaust fans by bringing fresh, clean air into the grow space, ensuring proper air exchange and circulation. Intake fans help maintain optimal temperature and CO_2 levels while preventing negative pressure buildup.

Intake fans are positioned near the bottom of the grow space to draw in fresh air from outside or from a designated intake area. They may include filters to remove contaminants and pests from incoming air.

3. Inline Fans:

Inline fans are versatile ventilation components used to facilitate air movement and ducting in hydroponic grow setups. These fans can be installed in ducting systems to exhaust air from grow lights, carbon filters, or other equipment.

Inline fans come in various sizes and configurations to accommodate different airflow requirements and ducting setups. They are often used in conjunction with carbon filters to remove odors and airborne contaminants from exhaust air.

4. Circulation fans:
Circulation fans are crucial in ensuring a consistent airflow and avoiding the formation of stagnant air zones in a grow area. They aid in the even distribution of air across the plant canopy, guaranteeing that each plant gets sufficient ventilation and CO_2.

5. Oscillating fans:
Oscillating fans are often utilized for their ability to create a gentle air movement by oscillating back and forth, which avoids causing excessive turbulence. To minimize hot spots and maintain consistent airflow throughout the entire canopy, it's crucial to position circulation fans strategically.

6. Controller systems:
Controller systems utilize automation to manage ventilation equipment, adjusting to specific parameters like temperature, humidity, and CO_2 concentration. Equipped with sensors, timers, and controllers, these systems continuously monitor the environment and modify fan speeds as needed. They ensure ideal growing conditions, reduce energy use, and lessen the need for manual control, granting cultivators enhanced command over their ventilation setups to cater precisely to plant requirements.

Incorporating specific ventilation components into hydroponic setups and employing effective ventilation strategies enables growers to foster a healthy and productive environment for cannabis cultivation. Maintaining consistent airflow, regulating temperature, and controlling humidity are crucial to optimize plant growth, prevent problems like mold and mildew, and ensure high-quality harvests.

CHAPTER 4

GERMINATION AND PROPAGATION

4.1 SEED SELECTION AND GERMINATION

Choosing Quality Seeds

Selecting high-quality seeds is essential to hydroponic marijuana cultivation since it lays the groundwork for a fruitful and profitable harvest. Vigorous development, maximum yields, and desired characteristics like taste, fragrance, and potency are all guaranteed by premium seeds. The following is a list of things to take into account while choosing high-quality seeds for hydroponic marijuana cultivation:

1. Genetics:
Begin by selecting reputable seed banks or breeders renowned for their high-quality genetics. Opt for well-established companies that have demonstrated consistent reliability and have garnered positive feedback from fellow growers.
Select seeds from stable, well-established genetic lines that exhibit desirable traits, including high potency, disease resistance, and uniform growth patterns. Investigate the genetic heritage of the strain to comprehend its origins and possible traits.

2. Seed Appearance:
Examine the seeds' physical characteristics to assess their quality. High-quality seeds usually have a dark hue, a firm texture, and a smooth, unbroken shell. Steer clear of seeds that appear light in

color, cracked, damaged, or underdeveloped, as these traits can indicate reduced viability and germination potential.

3. Viability and Germination Rate:
Selecting seeds with high viability and germination rates is crucial for a successful cultivation start. Opt for seeds that are fresh, have been stored correctly, and come from trustworthy suppliers to enhance the chances of successful germination. Many seed banks offer details on germination rates and seed viability, aiding you in making knowledgeable choices for your cultivation needs.

4. Strain Selection:
When selecting a strain for hydroponic cultivation, take into account your preferences, goals, and the specific growing conditions. Opt for a strain that matches your experience level, the space you have available, and the effects you desire, such as indica versus sativa or high CBD versus high THC. Investigate various strains to comprehend their growth habits, flowering periods, nutritional needs, and possible difficulties. Choose strains that are compatible with hydroponic growing techniques and meet your cultivation goals.

5. Customer Reviews and Recommendations:
Consult customer reviews and recommendations to gain insights from fellow cultivators who have grown the same strains. Search for feedback regarding yield, potency, flavor, aroma, and the general cultivation experience. Engage in online forums, social media groups, and community discussions to obtain advice, exchange experiences, and benefit from the successes and challenges of other growers.

6. Quality Assurance:
Select seeds from suppliers who emphasize quality assurance and testing to guarantee consistency, purity, and genetic integrity. Seek out seed banks that implement stringent quality control procedures, including genetic testing, seed certification, and germination guarantees. For ease and predictability, especially for novices or

those with limited experience in seed selection and cultivation, consider buying feminized or auto-flowering seeds.

By considering these factors and conducting thorough research, you can choose high quality seeds that meet your needs and set the stage for a successful hydroponic marijuana cultivation experience. Remember to invest in reputable suppliers, select strains with desirable traits, and prioritize viability, germination rates, and genetic integrity to maximize your chances of success.

Germination Techniques

Germination is the process where a seed starts to sprout and grow into a seedling. It is a vital initial step in hydroponic marijuana cultivation, setting the stage for a plant's healthy development. Below is an overview of typical germination methods employed in hydroponic marijuana cultivation:

1. Seed Selection:
Start with high-quality seeds from reputable suppliers to maximize germination success. Choose seeds that are dark in color, firm to the touch, and intact, with a smooth outer shell. Avoid seeds that are light-colored, damaged, or immature, as they may have lower viability and germination rates.

2. Pre-Soaking:
Pre-soaking seeds in water prior to planting can help kick-start the germination process by softening the seed coat and initiating hydration. Place seeds in a glass of room-temperature water and allow them to soak for 12-24 hours, ensuring they are fully submerged. Discard any seeds that float, as they may be less viable.

3. Usage of paper towels:
Using paper towels is a widely used method for seed germination as it provides a controlled environment. To start, moisten a paper towel with distilled water and distribute the seeds evenly on one half. Cover them by folding the other half of the towel over. Next, place

the paper towel containing the seeds into a plastic bag or container to maintain consistent humidity. Store the container in a warm, dark place, like atop a refrigerator or in a germination tray with a heating mat. Regularly check to ensure the paper towel stays moist but not soggy. Depending on the species and environmental conditions, seeds typically start to sprout within 2 to 7 days.

4. Direct Planting:

Many growers opt to sow seeds directly into their chosen substrate, such as rockwool cubes, coco coir, or perlite, bypassing pre-soaking or other germination methods. Simply create a small indentation in the substrate, deposit the seed, and gently cover it with more of the substrate.

To facilitate germination, ensure the substrate remains evenly moist but not saturated. It's also crucial to sustain warm conditions, ideally between 70-80°F (21-27°C), and to regulate humidity levels to support the growth of the seedlings.

5. Germination Trays or Domes:

Germination trays and domes create a controlled environment for seeds to germinate, equipped with humidity domes and heating mats to sustain ideal conditions. Sow seeds in separate cells or compartments containing a damp growing medium, and use a humidity dome to preserve moisture. Situate the germination tray or dome in an area that is warm and brightly lit with indirect sunlight. Regularly check the moisture and temperature to guarantee the best germination conditions.

6. Light and Temperature:

Provide consistent warmth and indirect light during the germination process to stimulate seedling growth. Optimal temperatures for germination typically range from 70-80°F (21-27°C). Avoid exposing germinating seeds to direct sunlight or extreme temperature fluctuations, as this can inhibit germination.

By using these germination techniques and providing optimal conditions, growers can maximize seed germination rates and establish healthy seedlings for successful hydroponic marijuana cultivation. Regular monitoring and adjustments to environmental factors such as temperature, humidity, and moisture levels are essential for promoting germination success and ensuring robust seedling development.

Transplanting Seedlings

Transplanting is the process of relocating young plants from their initial germination or seedling phase to a larger growth container or hydroponic system, where they will further grow and mature. This step is crucial in the hydroponic cannabis cultivation process. Employing proper transplanting techniques can minimize plant stress and promote robust root development. The summary of the transplanting process for hydroponic cannabis seedlings is as follows:

1. Timing:
Transplant seedlings once they have developed a strong root system and have at least two sets of true leaves. Avoid transplanting seedlings too early when they are still delicate and vulnerable to damage.

2. Prepare Growing Medium:
If transplanting into a soilless medium such as rockwool cubes, coco coir, or hydroponic grow medium, prepare the growing containers or trays beforehand. Ensure the medium is thoroughly moistened but not waterlogged to provide optimal conditions for root growth.

3. Gently Remove Seedlings
To gently remove seedlings, carefully extract them from their current container or growth medium, ensuring the roots remain intact. Hold the seedling by the stem and roots, and delicately loosen the surrounding medium to facilitate easy removal.

4. Handle Seedlings Carefully:

When handling seedlings, it's crucial to be gentle to prevent harm to the fragile stems, leaves, and roots. It's best to hold the seedlings by their cotyledons (seed leaves) or true leaves instead of the stems to avoid damage.

5.Transplanting Technique:
Make a hole or indentation in the fresh growing medium to fit the seedling's roots. Position the seedling so that its roots are evenly spread within the hole and not clumped together. Carefully add the medium around the roots, pressing it down softly to ensure the seedling is stable.

6.Watering:
After transplanting, it's important to water the seedlings thoroughly. This helps to settle the growing medium around the roots and ensures proper hydration. Be sure to use a gentle stream of water to prevent displacing the seedlings or causing soil erosion. The goal is to have the growing medium evenly moist, but not waterlogged.

7.Provide Optimal Conditions:
Place transplanted seedlings in a warm, well-lit area with indirect light to promote recovery and minimize stress. Maintain consistent environmental conditions, including temperature, humidity, and airflow, to support healthy growth and root development.

8.Monitor and Care:
Monitor transplanted seedlings closely in the days following transplanting to ensure they adjust well to their new environment. Keep an eye out for signs of stress, such as wilting or yellowing leaves, and make adjustments as needed to optimize growing conditions.

By following these transplanting guidelines and providing proper care and attention, growers can successfully transition seedlings into larger containers or hydroponic systems, setting the stage for robust growth and high yields in hydroponic marijuana cultivation. Regular monitoring and maintenance are essential to support seedling health and ensure a successful transition to the vegetative growth stage.

4.2 CLONING TECHNIQUES

Benefits Of Cloning

Cloning, or asexual propagation, is a popular technique used in hydroponic marijuana cultivation to reproduce genetically identical copies of a desirable mother plant. Cloning offers several benefits for growers, including:

1. Genetic Consistency:
Cloning allows growers to replicate the genetic traits of a high-performing mother plant, ensuring consistency in traits such as yield, potency, flavor, and aroma. This consistency is particularly valuable for cultivators aiming to produce a uniform crop with predictable characteristics.

2. Preservation of Desirable Traits:
Cloning from a chosen mother plant with favorable characteristics allows growers to sustain and replicate these traits over successive generations. This practice guarantees a consistent level of quality and performance, enabling the provision of a steady supply of exceptional genetics.

3. Time and Efficiency:
Cloning allows cultivators to avoid the time-consuming process of seed germination and phenotype selection. Clones can be rooted and grown much faster, conserving time and effort in the cultivation process. By cloning, cultivators can omit the vegetative growth phase and proceed straight to flowering, shortening the crop cycle and hastening the harvest time. This efficiency leads to more regular harvests and increased production turnover.

4. Cost Savings:
Cloning can be a cost-effective method of propagation compared to purchasing seeds or starter plants. Once established, mother plants can produce numerous clones without the need for additional seed purchases, saving money over time.

Additionally, by consistently producing high-quality clones from a known genetic source, growers can reduce the risk of crop failure or poor performance associated with unknown or inferior genetics.

5. Uniformity and Crop Management:
Cloning ensures uniformity in the crop, with all plants exhibiting identical growth characteristics and traits. This uniformity simplifies crop management, as growers can apply consistent cultivation practices, nutrient regimens, and environmental conditions across the entire crop.
Uniformity also facilitates efficient use of space and resources in hydroponic systems, as plants can be arranged and managed according to consistent growth patterns and requirements.

6. Pest and Disease Management:
Cloning from a healthy mother plant reduces the risk of introducing pests, diseases, or genetic abnormalities associated with seed propagation. Clones are genetically identical to the mother plant and inherit its resistance to pests, diseases, and environmental stressors. Additionally, cloning allows growers to maintain a stock of disease-free mother plants and backups, ensuring continuity of production and minimizing the impact of potential crop losses.

Overall, cloning offers numerous benefits for hydroponic marijuana cultivation, including genetic consistency, time and cost savings, enhanced crop management, and improved pest and disease management. By harnessing the power of cloning, growers can maximize efficiency, productivity, and quality in their cultivation operations.

Methods Of Cloning

Hydroponic marijuana cultivation offers various cloning methods, each with unique benefits tailored to the grower's preferences, resources, and level of expertise. Below is a summary of prevalent cloning techniques:

1. Cutting Method:

The cutting method is a popular cloning technique that involves taking cuttings from a healthy parent plant. Here's the standard procedure:

First, choose a vigorous parent plant with the desired characteristics and take cuttings from its lower branches or growing tips.

Next, with a sterile, sharp blade, make an angled cut just below a node. The cutting should be 4-6 inches long, with at least one set of nodes and leaves.

Remove any lower leaves from the cutting to minimize moisture loss and foster root growth.

Dip the cut end into rooting hormone to stimulate root formation, then insert it into a rooting medium like rockwool cubes, peat pellets, or coco coir.

Lastly, maintain suitable humidity, temperature, and lighting to encourage root development, usually in a propagation dome or tray with a heating mat and fluorescent or LED lights.

2. Aeroponic Cloning:

Aeroponic cloning involves suspending cuttings in an environment rich with nutrient mist or aerosol to encourage root growth. The typical process is as follows:

a) Set up a specialized aeroponic cloning system equipped with misters or sprayers to provide a consistent fine mist of nutrient solution to the cuttings.

b) Select cuttings from a vigorous mother plant following the cutting method, then place them into the aeroponic cloning system.

c) Maintain a consistent misting schedule and high humidity to foster root development.

d) Regularly check for root growth and transfer the cuttings with well-established roots to an appropriate growing medium.

3. Water Cloning:

Water cloning, also known as "cloning in water," involves rooting cuttings directly in water without a rooting medium. Here's how it's typically done:

Take cuttings from a healthy mother plant as described in the cutting method and place them in a container filled with clean, pH-balanced water.

Ensure the cuttings are submerged in water, with at least one set of nodes and leaves above the waterline.

Change the water regularly to prevent stagnation and maintain oxygen levels, typically every 2-3 days.

Monitor root development closely, and once roots are well-established, transplant rooted cuttings into a suitable growing medium.

4. Cube Method:

The cube method utilizes pre-soaked rockwool cubes or similar mediums for rooting cuttings. The typical process is as follows:

Soak rockwool cubes in pH-balanced water until they are fully saturated, then let the excess water drain.

Insert cuttings from a healthy mother plant into the pre-made holes in the rockwool cubes, as per the cutting method.

Place the cubes in a propagation tray or dome that maintains high humidity, and ensure they receive suitable lighting and temperature to promote root growth.

Regularly check the development of roots, and once they are well-established, move the rooted cuttings to a larger growing medium or into a hydroponic system.

Each cloning method offers its own set of benefits and challenges, allowing growers to select the one that aligns with their preferences, level of experience, and resources. Mastering these cloning techniques enables growers to produce healthy, genetically consistent plants, ensuring uniformity in their hydroponic cannabis cultivation.

Rooting Hormones and Solutions

Rooting hormones and solutions are crucial elements in hydroponic marijuana cultivation, particularly during the cloning process. They aid in the development of roots from cuttings, ensuring successful propagation and establishment of new plants. Here's a summary of rooting hormones and solutions commonly used in hydroponic cloning:

1.Types of Rooting Hormones:

Auxin-Based Hormones: These are the most commonly used rooting hormones in hydroponic cloning. Auxins, such as indole-3-butyric acid (IBA) and naphthaleneacetic acid (NAA), are plant hormones that stimulate root growth. They can be synthetic or naturally derived.

Gel Formulations: Rooting hormones commonly come in gel forms, offering a user-friendly approach for application to cuttings. The gel consistency ensures it sticks to the cut area, allowing for optimal contact and hormone uptake.

Liquid Formulations: Liquid rooting hormones are available as well, and they can be applied either by dipping the cutting's cut end into the solution or by misting the cutting with a spray bottle. Liquid formulations might have higher concentrations of active ingredients than gel forms.

2.Function of Rooting Hormones:

Rooting hormones work by stimulating cell division and growth in the cambium tissue of the cutting, encouraging the formation of roots. They help overcome the natural barriers to root development and promote the initiation of adventitious roots from the stem tissue. Rooting hormones also help protect the cuttings from fungal and bacterial pathogens by promoting faster root growth, which enhances the plant's ability to absorb water and nutrients from the growing medium.

3.Application of Rooting Hormones:

To apply rooting hormones, dip the cut end of the cutting into the hormone solution or gel, ensuring that the entire cut surface is coated. Alternatively, the cutting can be lightly misted or sprayed with the hormone solution.

It's essential to use rooting hormones according to the manufacturer's instructions and to avoid excessive application, as overdosing can inhibit root growth and cause damage to the cutting.

4.Rooting Solutions:

Besides rooting hormones, some growers also utilize rooting solutions or additives to further improve root development. These solutions often include vitamins, amino acids, and various nutrients that promote root growth and enhance the overall health of the plant. Rooting solutions can be mixed into the nutrient solution or applied directly to the root zone, offering additional nutrients and encouraging root growth, especially during the initial phases of rooting and transplantation.

Effective use of rooting hormones and solutions can enhance the cloning success rate for hydroponic growers, ensuring the robust establishment of new plants. Meticulous application and careful attention to detail are crucial for leveraging the full potential of these products and achieving successful propagation in hydroponic marijuana cultivation.

CHAPTER 5

VEGETATIVE STAGE: PROVIDING OPTIMAL CONDITIONS FOR GROWTH

Light Cycles

Light cycles are pivotal in hydroponic marijuana cultivation, affecting plant growth, flowering, and overall progression. Controlling light exposure's duration and intensity allows growers to steer cannabis plant growth stages and enhance yields. Below is a summary of light cycles typically employed in hydroponic marijuana cultivation.

1. Vegetative Stage:
During the vegetative stage, cannabis plants concentrate on growing leaves, stems, and branches. In hydroponic systems, this stage usually spans 4-8 weeks, varying with the targeted plant size and growth rate. Light cycles in this phase often involve 18-24 hours of light and 6-0 hours of darkness. Longer periods of light foster swift and robust vegetative growth, whereas shorter periods can induce more compact growth, useful for controlling plant height.

2. Flowering Stage
: The flowering stage of cannabis cultivation is initiated by altering the light cycle, usually to 12 hours of light and 12 hours of darkness. This shift cues the plants to move from vegetative growth to the flowering phase. Hydroponic growers can prompt this stage by setting the light schedule to 12 hours of continuous darkness

followed by 12 hours of light. It is vital to maintain complete darkness during the dark phase to ensure the flowers develop properly and the flowering cycle is not disrupted.

3.Transition Period:

Some cultivators opt for a transition phase between the vegetative and flowering stages to ready the plants for the shift in lighting schedule. Throughout this period, the lighting cycle is often shifted gradually from 18/6 to 12/12 over a span of one to two weeks. This transition phase is crucial for minimizing plant stress and facilitating a gradual adjustment to the new lighting regime, thereby diminishing the likelihood of shock and potential flowering initiation delays.

4.Light Spectrum:

In addition to controlling light cycles, growers may also manipulate the light spectrum to influence plant growth and development. During the vegetative stage, cannabis plants benefit from blue and white light, which promotes vegetative growth and leaf expansion. In the flowering stage, cannabis plants require a different light spectrum, with a focus on red and far-red light to stimulate flower development and resin production. Many hydroponic growers use LED grow lights with customizable spectra to provide the optimal light spectrum for each growth stage.

5.Light Intensity and Duration:

Throughout all growth stages, it's essential to maintain consistent light intensity and duration to ensure healthy plant growth and development. Light intensity should be adjusted according to the specific requirements of the cultivar, with higher intensity levels typically used during the vegetative stage and reduced intensity during flowering to mimic natural sunlight conditions.

Through meticulous management of light cycles in hydroponic marijuana cultivation, cultivators can encourage robust growth, enhance flowering, and attain peak yields. Consistent, precise, and detailed attention to light management is crucial for maximizing its potential as a vital element in cannabis cultivation.

Nutrients Requirements

Understanding nutrient requirements is crucial for hydroponic marijuana cultivation, as they significantly influence plant health, growth, and productivity. Cannabis plants need a balanced mix of macronutrients, micronutrients, and trace elements to flourish and yield high-quality products. Below is a summary of the nutrient needs for hydroponic marijuana cultivation:

1. Macronutrients:

Nitrogen is vital for the vegetative growth, leaf development, and overall vigor of plants. Cannabis plants, in particular, need higher nitrogen levels during the vegetative stage to support swift growth and biomass build-up. Nonetheless, an excess of nitrogen in the flowering stage can cause nutrient imbalances and adversely affect the development of flowers.

Phosphorus (P) is essential for root development, energy transfer, and flower formation. High levels of phosphorus are critical during the flowering phase to ensure robust bud development and resin production. Insufficient phosphorus can result in stunted growth and delayed flowering.

Potassium (K) is crucial for enzyme activation, water balance, and nutrient transportation in plants. It aids in the development of flowers, enhances quality, and bolsters overall plant health. A deficiency in potassium may result in subpar flower development and a heightened vulnerability to stress and diseases

Calcium (Ca) and Magnesium (Mg) are vital secondary macronutrients crucial for the health and growth of plants. They are significant in preserving cellular integrity, enzyme activation, and aiding photosynthesis. Insufficient calcium may result in blossom end rot among other physiological issues, whereas a deficiency in magnesium can lead to chlorosis and leaf discoloration.

2. Micronutrients

Iron (Fe) is essential for chlorophyll synthesis, photosynthesis, and enzyme functions. A deficiency in iron can lead to chlorosis, yellowing of leaves, and stunted plant growth.

Manganese (Mn) plays a role in photosynthesis, enzyme activation, and nitrogen metabolism. A deficiency in manganese can result in interveinal chlorosis and limited growth.

Zinc (Zn) is crucial for hormone regulation, enzyme activation, and protein synthesis. A lack of zinc can cause restricted growth and leaf deformities.

Copper (Cu) is important for enzyme activity, electron transport, and lignin formation. Copper deficiency can lead to leaf curling and wilting in plants.

Molybdenum (Mo) is vital for nitrogen fixation and enzymatic reactions. Insufficient molybdenum can lead to the yellowing of older leaves and impaired growth.

3.Trace Elements

Besides macronutrients and micronutrients, cannabis plants also need trace elements like boron (B), chlorine (Cl), and sulfur (S) in minor amounts for different physiological functions..

4.pH and Nutrient Absorption:

It is essential to keep the pH of the nutrient solution within the ideal range, usually between 5.5 and 6.5 for hydroponic systems, to ensure proper nutrient absorption and availability. Deviations from this pH range can impede nutrient uptake, potentially causing deficiencies or toxicities.

5.Nutrient Solution Management:

Hydroponic growers must regularly monitor and adjust the nutrient solution to ensure it meets the specific requirements of cannabis plants at each growth stage. This may involve using commercial nutrient formulations designed for hydroponic cultivation and following manufacturer recommendations for application rates and mixing instructions.

By providing cannabis plants with the appropriate balance of macronutrients, micronutrients, and trace elements in the nutrient solution, hydroponic growers can support healthy growth, optimize flower development, and achieve maximum yields. Regular monitoring, precise nutrient management, and attention to plant health are essential for success in hydroponic marijuana cultivation.

Pruning and Training Techniques

Pruning and training techniques are essential practices in hydroponic marijuana cultivation for managing plant growth, maximizing yields, and optimizing canopy efficiency. These techniques involve selectively removing or manipulating plant tissue to promote desirable growth patterns, increase light penetration, and improve airflow within the canopy. Here's a summary of common pruning and training techniques used in hydroponic marijuana cultivation:

1.Topping:
Topping is a horticultural technique that involves cutting off the apical meristem, or the topmost growth point, of the main stem, usually during the vegetative phase. This method encourages the plant to produce several main colas or flowering sites, rather than just one dominant cola. Topping the plant allows growers to encourage lateral growth, resulting in a more balanced canopy with numerous tops, which optimizes light absorption and leads to more consistent bud growth.

2.FIMing:
FIMing (short for "f*ck, I missed") is a variation of topping that involves removing a portion of the apical meristem rather than the entire tip. This results in a less pronounced removal of growth, often leading to four or more new growth tips instead of two.
FIMing can be a less stressful alternative to topping and may produce bushier plants with multiple tops and increased bud sites.

3. LST (Low Stress Training):
LST is a training technique used to manipulate the shape and structure of the plant by gently bending and securing stems or branches. This encourages lateral growth and creates a more horizontal canopy, maximizing light exposure to lower bud sites.
To perform LST, gently bend stems or branches horizontally and secure them in place using soft ties or plant clips. Gradually increase tension over time to encourage gradual bending and prevent damage to the plant tissue.

4. SCROG (Screen of Green):
SCROG, or Screen of Green, is a cultivation method that uses a horizontal screen or net to guide and support a plant's canopy. The screen is placed above the plants, and as they grow, their stems and branches are woven horizontally across the screen. This technique fosters an even canopy with numerous buds at the same level, enhancing light exposure and encouraging consistent bud growth. Additionally, it aids in controlling plant height and optimizing space in the growing area.

5. Defoliation:
Defoliation involves selectively removing leaves from the lower and inner parts of the plant canopy to improve airflow, reduce humidity, and redirect energy to upper bud sites. This practice can help prevent mold and mildew issues and increase light penetration to lower bud sites.
Defoliation is typically performed during the vegetative stage and early flowering stage, avoiding excessive leaf removal during late flowering to ensure sufficient energy production for bud development.

6. Super-cropping:
Super-cropping is a method that involves stressing the plant by gently squeezing or bending the stems to partially break the internal tissues. This action temporarily disrupts the flow of nutrients and water, encouraging the plant to develop thicker stems and more robust branches. Super-cropping is beneficial for controlling plant

height, enhancing canopy architecture, and boosting bud yield. However, it must be done with caution to prevent excessive damage to the plant tissues.

7. Pruning Suckers and Lower Growth:
Removing suckers (small shoots that develop in the crotch between stems and branches) and lower growth helps redirect energy to top bud sites and promotes better airflow and light penetration within the canopy. Pruning these lower branches and growth helps focus plant energy on producing larger, more potent buds in the upper canopy. By implementing these pruning and training techniques in hydroponic marijuana cultivation, growers can manage plant growth, optimize canopy structure, and maximize yields. It's essential to tailor these techniques to the specific needs of each cultivar and growth stage, and to monitor plant health closely to ensure optimal results.

5.1 PREVENTING PEST AND DISEASES

Hydroponic marijuana growers must be vigilant against common pests and diseases that can threaten plant health and reduce yield. Although hydroponic systems may mitigate some risks associated with soil-based cultivation, pests and diseases can still occur. Below is a list of common pests and diseases found in hydroponic marijuana growing:

Common Pests

1. Spider Mites:
Spider mites, small arachnids that consume plant sap, can cause stippling, leaf yellowing, and necrosis. Commonly located on the undersides of leaves, they reproduce rapidly in warm and dry environments.

2. Aphids:

Aphids, tiny insects with soft bodies, feed on plant sap and secrete honeydew. This secretion can cause leaves to curl, stunt plant growth, and promote the growth of sooty mold. They have a rapid reproduction rate and can quickly infest plants in great numbers.

3.Whiteflies:
Whiteflies, resembling tiny moths, consume plant sap and deposit their eggs beneath the leaves. Severe infestations may result in yellowed leaves, wilting, and early leaf fall, which can diminish plant health and productivity.

4.Thrips:
Thrips are minuscule, slender insects that nourish themselves by piercing plant cells and extracting their contents. These pests can lead to stippling, silvering, and distortion of leaves, in addition to transmitting viruses that are harmful to plants.

5.Fungus Gnats:
Fungus gnats are tiny flies that deposit their eggs in damp growing mediums. The resulting larvae consume organic material and plant roots, causing root harm, nutrient shortages, and plant wilting. The adult gnats, meanwhile, are bothersome pests that can spread contamination to plants and their growing environments.

Common Diseases

1.Powdery Mildew:
Powdery mildew is a fungal infection marked by a white, powdery residue on leaves, stems, and flower buds. It flourishes in warm, damp environments and can quickly proliferate, causing leaf deterioration, diminished photosynthesis, and decreased agricultural production.

2.Botrytis (Gray Mold):
Botrytis is a fungal disease characterized by grayish-brown mold on plant tissues, often occurring in areas with high humidity and inadequate air circulation. It affects flowers, buds, and stems, resulting in bud rot and significant crop losses.

3. Pythias (Root Rot):
Pythias, a waterborne fungus, infects plant roots, leading to rot and wilting. It prospers in wet and inadequately drained environments, which results in diminished nutrient absorption and hindered growth.

4. Fusarium:
Fusarium, a soilborne fungus, attacks plant roots and vascular tissues, leading to wilting and yellowing. This infection can cause seedlings to damp off and mature plants to decline.

5. Verticillium:
Verticillium is a soilborne fungus that infects plant vascular tissues, causing wilting and yellowing. It can lead to reduced nutrient uptake and stunted growth.

Integrated Pest Management (IPM) Strategies

Integrated Pest Management (IPM) is an effective and environmentally sustainable approach to managing pests in hydroponic marijuana cultivation. It combines multiple strategies to prevent, monitor, and control pests while minimizing the use of chemical pesticides. Here's a summary of common IPM strategies used in hydroponic cultivation:

1. Cultural Controls:

Sanitation: It is crucial to maintain clean and hygienic conditions to avert pest infestations. Regular removal of plant debris, fallen leaves, and any dead or diseased plants is key to reducing potential habitats for pests.

Quarantine: Before integrating new plants and growth mediums into your hydroponic system, inspect them for pests or diseases. Isolating new additions temporarily can help prevent the introduction and spread of pests.

Crop Rotation: Disrupting pest life cycles and diminishing their environmental presence can be achieved by rotating crops or altering planting sites between cycles.

Optimal Growing Conditions: Ensuring plants have ideal conditions, such as proper lighting, temperature, humidity, and nutrients, is essential to foster robust growth and bolster their pest and disease resistance.

2. Biological Controls:
Introducing Predatory Insects: Release beneficial insects like ladybugs, lacewings, predatory mites, and parasitic wasps into the environment to naturally reduce pest populations. These predators can effectively manage aphids, spider mites, thrips, and other harmful pests, eliminating the need for chemical pesticides.
Utilizing Microbial Inoculants: Introduce beneficial microorganisms, such as Bacillus thuringiensis (Bt) or Beauveria bassiana, into the soil to curb pest populations. These targeted microbes are adept at controlling both larvae and adult stages of specific insect pests.

3. Mechanical and Physical Controls:
Traps and Barriers: Employ sticky traps, pheromone traps, and yellow sticky cards to monitor and capture flying insects like fungus gnats, whiteflies, and thrips. Set up physical barriers, such as screens or nets, to block pests from accessing the growth area.

Pruning and Removal: Excise and dispose of infested plant components, such as leaves, stems, and flowers, to inhibit the proliferation of pests and diseases throughout the crop.

4. Chemical Controls (as a last resort):
When non-chemical methods fail to control pest populations effectively, it may be necessary to resort to selective and minimally toxic pesticides. Opt for products that are specifically approved for hydroponic systems and adhere strictly to the label instructions to reduce risks to human health and the environment. Steer clear of broad-spectrum pesticides, as they can harm beneficial insects and upset the ecosystem's natural equilibrium.

5. Monitoring and Record-Keeping:
Regularly monitor plants for signs of pest infestations, including damage to leaves, wilting, discoloration, and presence of pests or

their eggs. Keep detailed records of pest sightings, control measures implemented, and their effectiveness to inform future pest management decisions.

By integrating these IPM strategies into hydroponic marijuana cultivation, growers can effectively manage pest populations while minimizing reliance on chemical pesticides and promoting a healthy and sustainable growing environment. Regular monitoring, proactive prevention, and careful implementation of control measures are essential for successful pest management in hydroponic systems.

Organic Pest Control Methods

Organic pest control is vital for a healthy and sustainable hydroponic marijuana cultivation system. It emphasizes natural, non-toxic solutions to control pests, thereby reducing environmental damage, preserving beneficial insects, and safeguarding human health. Below are some effective organic pest control strategies appropriate for hydroponic setups:

1. Biological Control:

Beneficial Insects: Employing predatory insects like ladybugs, lacewings, predatory mites, and parasitic wasps can naturally curb pest populations. These beneficial predators are effective in managing pests such as aphids, spider mites, and thrips, reducing the reliance on chemical pesticides.

Nematodes: Introducing beneficial nematodes into the soil can effectively control pest populations, including fungus gnats, root aphids, and root mealybugs. These tiny predators hunt and eliminate larvae, leading to a gradual reduction in pest numbers.

2. Botanical Insecticides:

Neem Oil: Extracted from the neem tree, serves as a natural insecticide, repellent, and growth regulator. It disrupts the feeding, reproduction, and development of several pests, including aphids, whiteflies, and spider mites, yet it is relatively safe for beneficial insects and mammals.

Pyrethrin: Derived from chrysanthemum flowers, is a natural insecticide effective against various pests. It interferes with their nervous systems, controlling problems caused by aphids, thrips, and whiteflies. Pyrethrin degrades rapidly in the environment and is deemed safe for organic farming when used as directed.

3. Microbial Pesticides:

Bacillus thuringiens:this is, commonly referred to as Bt, is a bacterium found in nature that generates proteins harmful to specific insect larvae such as caterpillars and mosquitoes. It can be applied directly onto plants or mixed into the soil. There are different formulations of Bt that are tailored to target particular pests. Beauveria bassiana is a fungal pathogen that infects and rapidly exterminates various insects. It proves effective against a multitude of pests, including thrips, aphids, and white flies. This fungus can be utilized to manage insect populations through methods such as foliar sprays or soil drenches.

4. Organic Sprays and Solutions:

Garlic and Pepper Sprays: Sprays made from garlic and pepper can deter and repel pests such as aphids, spider mites, and whiteflies. These homemade solutions can be directly applied to the foliage of plants to discourage feeding and interrupt the behavior of pests. Soap and Oil Sprays: Sprays containing soap and oil, like insecticidal soap and horticultural oil, work by suffocating and disrupting the cell membranes of soft-bodied insects, including aphids, mites, and thrips. These sprays are effective in controlling minor pest outbreaks and are generally safe for use on a wide range of plants.

5. Companion Planting and Cultural Practices:

Companion Planting: Planting pest-repelling herbs and flowers like basil, marigold, and mint among cannabis plants can help deter pests and attract beneficial insects. Companion planting can also improve biodiversity and create a more resilient ecosystem.

Crop Rotation and Diversity: Rotate crops or interplant different varieties of cannabis to disrupt pest life cycles and reduce pest pressure over time. Diversity in plant species and genetics can also make it harder for pests to establish and spread.

Incorporating organic pest control methods into hydroponic marijuana cultivation allows growers to effectively manage pests and promote a sustainable, healthy growing environment. Success in pest management within organic hydroponic systems hinges on regular monitoring, maintaining proper sanitation, and implementing proactive prevention strategies.

CHAPTER 6

FLOWERING STAGE

6.1 TRANSITIONING TO FLOWERING

Light Schedule Adjustments

Light schedule adjustments are an important aspect of hydroponic marijuana cultivation, especially during different growth stages. By manipulating the duration and timing of light exposure, growers can influence plant growth, flowering, and overall development. Here are some common light schedule adjustments for hydroponic marijuana cultivation:

1. Vegetative Stage:
During the vegetative stage, cannabis plants require longer periods of light to promote vigorous vegetative growth and leaf development. A typical light schedule for the vegetative stage is 18-24 hours of light followed by 6-0 hours of darkness.
Growers may choose to provide continuous light (24 hours) during the vegetative stage to maximize growth rates and shorten the overall vegetative period. However, some growers prefer to provide a dark period to allow plants to rest and conserve energy.

2. Transition to Flowering:
As growers shift from the vegetative to the flowering stage, it's necessary to modify the lighting schedule to encourage blooming. Usually, this means cutting back the light from 18-24 hours to a balanced 12 hours of light and 12 hours of continuous darkness. It's

best to ease plants into this new schedule over a week or two, reducing stress and promoting a seamless change. This steady approach allows plants to adapt to their new lighting regime smoothly, avoiding any shock.

3. Flowering Stage:
During the flowering phase, cannabis plants need a consistent cycle of 12 hours of light followed by 12 hours of darkness to trigger and sustain the development of flowers. It's vital to adhere to a strict lighting schedule during this phase to avoid any disruptions in the flowering process. It is important to ensure that the plants are in total darkness during their dark cycle to prevent light pollution, which can interfere with the flowering cycle and cause inconsistent growth or hermaphroditism.

4. Light Intensity and Spectrum:
In addition to adjusting the light schedule, growers should also pay attention to light intensity and spectrum. Providing the correct intensity and spectrum of light can optimize plant growth, resin production, and overall yield.

Use high-quality grow lights with adjustable intensity and customizable spectra to meet the specific light requirements of cannabis plants at each growth stage. Consider using full-spectrum LEDs or high-pressure sodium (HPS) lights for optimal results.

5. Monitoring and Adjustment:
Regularly monitor plant response to the light schedule and make adjustments as needed based on plant growth and development. Pay attention to signs of stress, such as leaf curling, yellowing, or stretching, which may indicate inadequate light exposure.

Keep a log of light schedules and plant responses to track progress and make informed decisions about future adjustments. Experiment with different light schedules and observe how plants respond to optimize growth and flowering.

By carefully adjusting the light schedule throughout the different growth stages, hydroponic growers can promote healthy growth, maximize yields, and achieve optimal results in marijuana cultivation. Consistency, attention to detail, and responsiveness to plant needs are key to successful light management in hydroponic systems.

Nutrient Adjustments

Adjusting nutrients is vital for maintaining the ideal nutrient levels and pH balance in hydroponic cannabis cultivation. Effective nutrient management guarantees that the plants get the necessary elements required for robust growth and development. The following are important factors to consider when adjusting nutrients in hydroponic setups:

1.Nutrient Solution Concentration:
Monitor the concentration of nutrients in the hydroponic solution using an electrical conductivity (EC) meter or a total dissolved solids (TDS) meter. Adjust the nutrient solution strength as needed to maintain the desired EC or TDS levels.

During the vegetative stage, plants typically require higher nutrient concentrations to support rapid growth. Gradually increase nutrient levels as plants mature and adjust according to plant response. Avoid overfeeding plants with excessive nutrient concentrations, as this can lead to nutrient imbalances, nutrient lockout, and nutrient toxicity. Regularly check and adjust nutrient levels to prevent overfeeding.

2.pH Adjustment:
Regular monitoring and adjustment of the nutrient solution's pH is crucial for optimal nutrient uptake. The preferred pH range for hydroponic cannabis cultivation is generally between 5.5 and 6.5. A pH meter or test kit can be used to assess the nutrient solution's pH levels. If the pH strays from the desired range, it can be adjusted using pH up or pH down solutions, like potassium hydroxide or phosphoric acid.

Variations in pH levels can result from factors such as plant nutrient absorption, water quality, and microbial processes in the growth medium. To maintain stability, it's essential to consistently monitor and adjust pH levels when needed.

3. Nutrient Ratio Adjustments:

Ensure that the nutrient solution provides a balanced ratio of essential macro and micronutrients for optimal plant growth. Adjust nutrient formulations or supplements as needed to address specific nutrient deficiencies or imbalances.

Common macronutrients include nitrogen (N), phosphorus (P), potassium (K), calcium (Ca), and magnesium (Mg). Micronutrients such as iron (Fe), manganese (Mn), zinc (Zn), copper (Cu), and boron (B) are also essential for plant health.

Use commercial hydroponic nutrient formulations or create custom nutrient solutions using individual fertilizer salts to tailor nutrient ratios to the specific needs of cannabis plants at each growth stage.

4. Nutrient Solution Temperature:

Monitoring the temperature of the nutrient solution is crucial to keep it within the optimal range for nutrient absorption and microbial activity. The ideal temperature for a nutrient solution is usually between 18°C and 22°C (65°F to 72°F). Regularly measure the nutrient solution temperature with a water thermometer. Should the temperature rise above the desired range, think about employing a water chiller or modifying the grow room's ambient temperature to achieve cooler conditions.

5. Plant Response Monitoring:

Pay close attention to plant response to nutrient adjustments, including leaf color, growth rate, and overall health. Adjustments may be necessary based on observed symptoms of nutrient deficiencies or toxicities.

Keep a detailed record of nutrient adjustments, plant responses, and environmental conditions to track progress and make informed decisions about future adjustments. Experiment with different nutrient formulations and strategies to optimize plant health and maximize yields.

By carefully monitoring and adjusting nutrient levels, pH, and other factors in the hydroponic system, growers can provide cannabis plants with the essential elements they need for healthy growth and development. Consistent monitoring, attention to detail, and responsiveness to plant needs are key to successful nutrient management in hydroponic marijuana cultivation.

Environmental Considerations

Environmental considerations are critical for successful hydroponic marijuana cultivation, as they directly impact plant health, growth, and overall yield. Here are key environmental factors to consider:

1.Temperature:

Maintain optimal temperature conditions for cannabis growth, typically between 70-85°F (21-29°C) during the day and slightly cooler at night. Avoid temperature fluctuations, as they can stress plants and affect growth.

Use heating or cooling systems to regulate temperature as needed, especially in indoor grow environments where external temperatures may fluctuate.

2.Humidity:

Managing humidity levels is crucial to prevent mold, mildew, and other moisture-related problems. The ideal humidity levels differ by the plant's growth stage: approximately 60-70% during the vegetative phase and 40-50% during the flowering stage. Employ dehumidifiers to lower excessive humidity, and use humidifiers to raise it when too low, particularly in arid climates or indoor growing environments..

3.Air Circulation:

To prevent stagnant air and maintain appropriate CO_2 levels, ensure sufficient air circulation. Utilize oscillating fans to enhance airflow within the grow space, aiding in mold prevention and fortifying plant stems. Set up intake and exhaust fans to introduce fresh air and expel stale air, preserving a steady flow throughout the grow area.

4. CO2 Levels:

Supplement CO_2 levels in the grow room to enhance plant growth during the vegetative stage. CO_2 levels of 1000-1500 ppm (parts per million) are optimal for promoting photosynthesis and maximizing growth.

Use CO_2 generators or tanks with regulators to maintain consistent CO_2 levels, and ensure proper ventilation to prevent CO_2 buildup.

5. Lighting:

Ensure all plants receive adequate and uniform lighting that simulates natural sunlight. Utilize high-quality grow lights with the correct spectrum for each stage of growth: blue for vegetative growth and red for flowering. Keep light cycles consistent, with 18-24 hours of light for vegetative growth and a 12-hour light/dark cycle during the flowering phase.

6. Nutrient Solution Temperature:

It is crucial to keep the temperature of the nutrient solution within the ideal range of 65-75°F (18-24°C). Lower temperatures can enhance oxygen solubility and nutrient absorption, whereas higher temperatures can encourage microbial growth. For temperature control, particularly in recirculating systems prone to variations, employ water chillers or heaters as necessary.

7. pH and EC Levels:

Consistent monitoring and adjustment of pH and electrical conductivity (EC) levels in the nutrient solution are essential for the optimal uptake of nutrients. Typically, the ideal pH range should be between 5.5 and 6.5, and the EC levels must be adjusted to meet the particular growth phase and nutritional requirements of the plants.

8. Environmental Monitoring:

Continuously monitor environmental conditions using sensors and meters for temperature, humidity, CO_2 levels, and light intensity. Keep detailed records and adjust environmental parameters as needed to maintain optimal growing conditions.

By carefully managing environmental factors in hydroponic marijuana cultivation, growers can create an ideal growing environment that promotes healthy plant growth, maximizes yields, and minimizes the risk of pests, diseases, and environmental stressors. Regular monitoring and adjustments are essential for maintaining stable and optimal conditions throughout the entire growth cycle.

6.2 MAXIMIZING FLOWERING AND BUD DEVELOPMENT

Supplemental Lighting Techniques

Supplemental lighting techniques are essential in hydroponic marijuana cultivation to ensure plants receive adequate light intensity and spectrum for optimal growth and development, especially in indoor or greenhouse settings where natural sunlight may be limited or inconsistent. Here are some supplemental lighting techniques commonly used in hydroponic cultivation:

1. Light Emitting Diode (LED) Supplemental Lighting:
LED lights are energy-efficient and offer high customization, enabling growers to adjust the light spectra to meet the precise needs of cannabis plants during various stages of growth. Supplemental LED lighting can deliver extra light intensity and spectrum, which can improve plant growth and flowering.
Utilize LED grow lights featuring adjustable spectrums encompassing both blue and red wavelengths to cater to vegetative growth and flowering stages. Advanced LED fixtures may also provide full-spectrum lighting, closely simulating natural sunlight. Strategically place supplemental LED lights to achieve even light distribution and coverage throughout the canopy, effectively complementing natural or primary light sources.

2. High-Pressure Sodium (HPS) or Metal Halide (MH) Supplemental Lighting:

HPS and MH lights are traditional options for supplemental lighting in hydroponic cultivation. MH lights emit a blue spectrum ideal for vegetative growth, while HPS lights produce a red spectrum suitable for flowering.

Supplemental HPS or MH lighting can be used in conjunction with natural sunlight or primary LED fixtures to boost light intensity and provide additional spectral coverage, especially during periods of low natural light or short days.

Adjust the distance and positioning of HPS or MH lights to achieve optimal light intensity and coverage without causing heat stress or light burn to plants.

3. Light Movers:

Light movers are devices that transport grow lights across a track or rail system, ensuring even light distribution over the plant canopy. They guarantee uniform light exposure to all parts of the plant, enhancing light penetration and photosynthetic efficiency.

Employing light movers alongside primary or additional lighting fixtures can optimize light spread and reduce shading within the grow area, leading to more consistent plant development and increased yields.

4. Lighting Controllers and Timers:

Utilize lighting controllers and timers to automate supplemental lighting schedules and ensure consistent light cycles throughout the growth stages. Set timers to synchronize supplemental lighting with natural sunlight or primary lighting sources, adjusting as needed based on seasonal changes or plant requirements.

Programmable controllers allow growers to customize light intensity, spectrum, and duration, optimizing supplemental lighting strategies for maximum plant health and productivity.

5. Light Spectrum Adjustments:

Experiment with different light spectra and supplemental lighting combinations to determine the most effective lighting strategy for cannabis plants at each growth stage. Tailor light spectra to promote specific physiological responses such as vegetative growth, flowering, or resin production.

Consider using supplemental UVB or far-red light wavelengths to enhance secondary metabolite production and improve cannabinoid and terpene profiles in cannabis flowers, potentially enhancing medicinal or recreational properties.

6. Light Stress Management:

Monitor plants closely for signs of light stress, including leaf bleaching, leaf curling, or stunted growth, and adjust supplemental lighting intensity or duration accordingly to avoid phototoxicity or light-induced stress.

Gradually introduce supplemental lighting to acclimate plants to increased light levels, especially during transitions between growth stages or when implementing new lighting strategies.

By implementing these supplemental lighting techniques in hydroponic marijuana cultivation, growers can optimize light conditions, promote healthy plant growth, and maximize yields throughout the entire growth cycle. Experimentation, observation, and careful adjustment are key to developing effective supplemental lighting strategies tailored to specific growing environments and plant requirements.

Flowering Boosters and Enhancers

Flowering boosters and enhancers are supplements used in hydroponic marijuana cultivation to stimulate flower production, improve flower quality, and increase overall yield during the flowering stage. These products typically contain a blend of nutrients, vitamins, hormones, and natural extracts designed to promote flowering and enhance floral development. Here are some common types of flowering boosters and enhancers used by hydroponic growers:

1. Phosphorus and Potassium Supplements:

Phosphorus (P) and potassium (K) are critical macro nutrients necessary for the development of flowers and fruiting in cannabis plants. Enriching flowering boosters with phosphorus and potassium supplies the additional nutrients required for vigorous flower growth and bud development. During the flowering stage, seek nutrient formulations with a higher proportion of phosphorus and potassium relative to nitrogen (N). These supplements, commonly referred to as bloom boosters or bloom enhancers, are designed expressly for the flowering phase.

2. Bud Density and Size Enhancers:

Some flowering boosters are designed to increase bud density, size, and weight by providing additional nutrients and hormonal support. These products typically contain amino acids, sugars, and natural extracts that promote cell division, flower elongation, and resin production.

Ingredients such as humic acids, kelp extract, and molasses are commonly used in bud density enhancers to improve nutrient uptake, enhance flavor, and increase terpene production in cannabis flowers.

3. PK Boosters:

PK boosters are concentrated formulations containing high levels of phosphorus (P) and potassium (K) designed to maximize flower production and quality. These products are often used during the peak flowering phase to provide an extra boost of essential nutrients when plants have increased nutrient demands.

PK boosters are available in various formulations and strengths, allowing growers to adjust nutrient levels based on plant response and specific strain requirements. They can be used in combination with base nutrients or as standalone supplements during the flowering stage.

4. Flower Stimulators and Hormonal Supplements:

Flower stimulators contain natural plant hormones and growth regulators that promote flowering initiation, accelerate flower development, and increase flower yield. These products may contain ingredients such as cytokinins, gibberellins, and auxins that stimulate hormonal responses in cannabis plants.

Hormonal supplements can help synchronize flowering across different plants, improve flower uniformity, and shorten flowering times, leading to more consistent harvests and higher yields.

5. Terpene and Essential Oil Enhancers:

Terpene and essential oil enhancers aim to increase the production of aromatic compounds, terpenes, and essential oils in cannabis flowers. These enhancers typically include amino acids, vitamins, and botanical extracts to boost terpene synthesis and accumulation.

Higher terpene levels can improve the aroma, flavor, and therapeutic benefits of cannabis flowers, thus increasing their appeal to consumers. Growers who strive to cultivate aromatic, high-quality cannabis strains frequently use terpene enhancers.

6. Organic and Bio-Stimulant Enhancers:

Organic flowering enhancers contain natural ingredients derived from organic sources, such as composted plant materials, seaweed extracts, and microbial inoculants. These bio-stimulants promote soil health, microbial activity, and nutrient cycling, resulting in improved flower quality and yield.

Organic enhancers are suitable for growers practicing organic or sustainable cultivation methods and are environmentally friendly alternatives to synthetic fertilizers and chemical additives.

When using flowering boosters and enhancers in hydroponic marijuana cultivation, it's essential to follow manufacturer recommendations and dosage instructions carefully to avoid overfeeding or nutrient imbalances. Experimentation may be necessary to determine the most effective products and application methods for specific strains and growing conditions. Additionally, growers should monitor plant response closely and adjust nutrient supplementation based on observed results and plant needs.

Harvest and Timing Techniques

The timing and techniques of harvesting play crucial roles in hydroponic marijuana cultivation, directly affecting the final product's potency, quality, and yield. Ensuring plants are harvested at the optimal time and using proper methods guarantees the realization of their full potential, yielding top-quality buds. Below is a guide on the methods and timing for harvesting hydroponic marijuana:

1. Determining Harvest Readiness:

Monitoring trichomes is crucial: Trichomes, the small, mushroom-shaped glands on cannabis flowers, house cannabinoids and terpenes. To assess them, employ a magnifying tool like a jeweler's loupe or a digital microscope, and observe any shifts in their color and clarity..

Harvest when trichomes turn milky white: Trichomes shift from clear to milky white during the maturation process in the flowering stage. Harvesting at a time when the trichomes are mostly milky white is indicative of the highest levels of cannabinoids and terpenes, ensuring optimal potency and flavor.

Account for strain-specific traits: Various cannabis strains display distinct flowering behaviors and rates of maturation. Certain strains might take more time to flower, resulting in denser buds, whereas others could mature faster, yielding flowers with a less compact structure.

2. Flushing:

Flushing refers to the practice of watering plants with pH-balanced pure water during the last days or weeks before harvesting. This process removes excess nutrients and mineral buildup from both the growing medium and the plant tissues. Flushing enhances the quality and taste of the final product by minimizing nutrient accumulation, resulting in a smoother and cleaner smoke. Additionally, it enables plants to use up the remaining nutrients and compounds, which contributes to a cleaner burn and finer ash.

3. Harvesting Techniques:

Utilize sharp, sanitized scissors or pruning shears for cutting branches: Sever branches one at a time instead of harvesting the whole plant simultaneously. This method simplifies the handling and processing of the material.

Make cuts at the stem's base: Execute a clean, angled cut at the branch's main stem to reduce harm to the plant and aid in its drying and curing process.

Eliminate large fan leaves and unnecessary foliage: Remove the bulky fan leaves and surplus foliage from the branches to enhance air circulation and support uniform drying. Preserve the smaller sugar leaves as they are rich in trichomes and add to the final product's potency.

Treat harvested materials with care: Refrain from squeezing or compressing the buds to avoid harming the trichomes and maintain the flower's structure.

Dry branches by hanging them upside down: Place the harvested branches in an inverted position in a cool, dark, and well-ventilated area to dry, ensuring the relative humidity remains between 50-60%. Employ drying racks or lines to hang the branches, permitting airflow around each bud.

4. Drying and Curing:

Dry harvested buds slowly: Allow harvested buds to dry slowly over 7-14 days, depending on environmental conditions and plant characteristics. Slow drying helps preserve terpenes and cannabinoids while reducing the risk of mold and mildew.

Monitor drying conditions: Check drying buds regularly for moisture levels and signs of mold or mildew. Maintain proper humidity and temperature levels in the drying area to ensure a smooth and controlled drying process.

Cure dried buds in glass jars: After drying, transfer buds to glass jars for the curing process. Seal jars tightly and store them in a cool, dark place with a humidity level of around 60-65%. Open jars periodically to release excess moisture and allow air exchange. Cure buds for 2-4 weeks: Cure buds for 2-4 weeks to allow flavors and aromas to develop fully and to improve the smoothness of the smoke. Proper curing enhances the overall quality, potency, and shelf life of the final product.

By following these harvest timing and techniques in hydroponic marijuana cultivation, growers can ensure optimal potency, flavor, and quality in their harvested buds. Proper timing, handling, and curing are essential steps to maximize the value and enjoyment of the final product.

CHAPTER 7

HARVESTING AND CURING

7:1 HARVESTING YOUR CROP

Signs of Maturity

Signs of maturity in hydroponic marijuana plants indicate that they are ready for harvest. Monitoring these signs is crucial for determining the optimal time to harvest and achieve the desired potency, flavor, and yield. Here are some common signs of maturity to look for:

1. Trichome Development:
Trichomes are tiny, mushroom-shaped glands on the surface of cannabis flowers that contain cannabinoids and terpenes. They are an essential indicator of maturity and potency.
Monitor the color and clarity of trichomes using a magnifying tool such as a jeweler's loupe or a digital microscope. Mature trichomes typically change from clear to cloudy or milky white, indicating peak cannabinoid content.

Some growers may also look for amber-colored trichomes, which indicate a higher level of cannabinoid degradation and may be preferred for certain effects, such as relaxation or sedation.

2. Pistil Color:

Pistils, the hair-like structures emerging from the calyxes of female cannabis flowers, serve as indicators for harvest timing by changing color as the plant matures. They start off white or light-colored at the onset of flowering and transition to hues of orange, brown, or red, varying with the strain. While a shift from white to a darker shade in pistils is a widely recognized sign of maturity, it's best to consider it alongside other maturity markers, such as the state of trichome development.

3. Bud Size and Density:

Mature cannabis buds typically reach their full size and density towards the end of the flowering stage. They may appear plump, dense, and tightly packed with resinous trichomes.

Monitor bud development throughout the flowering phase, paying attention to size, density, and overall structure. Buds that are large, dense, and resinous are indicative of maturity and high-quality flowers.

4. Flower Structure:

Cannabis flowers develop a characteristic structure as they mature, with dense clusters of calyxes and pistils forming along the stems and branches.

Mature flowers may exhibit swollen calyxes, swollen resin glands, and a frosty appearance due to the abundance of resinous trichomes covering the surface.

5. Aroma and Terpene Profile;

Aroma and terpene profile are important indicators of maturity and flavor development in cannabis flowers. As plants mature, they produce a complex blend of aromatic compounds known as terpenes. Monitor the aroma of flowering plants as they approach maturity, noting changes in scent and intensity. Mature flowers may emit strong, distinctive aromas characteristic of the strain's terpene profile.

6. Overall Plant Health:

Healthy, vigorous plants are more likely to reach maturity and produce high-quality flowers. Monitor overall plant health, including leaf color, texture, and growth patterns, throughout the flowering phase.

Address any issues such as nutrient deficiencies, pests, or diseases promptly to ensure optimal plant health and minimize stress during the final stages of flowering.

By observing these signs of maturity in hydroponic marijuana plants, growers can determine the optimal time to harvest and achieve the desired potency, flavor, and yield in their harvested buds. It's essential to monitor multiple indicators and use a combination of visual, olfactory, and tactile cues to assess maturity accurately.

Harvesting Techniques

Harvesting techniques play a crucial role in hydroponic marijuana cultivation, as they determine the quality, potency, and overall yield of the harvested buds. Proper harvesting techniques ensure that plants are harvested at the peak of maturity and handled carefully to preserve their integrity. Here's a guide to essential harvesting techniques for hydroponic marijuana:

1. Timing:

The timing of the harvest is crucial for maximizing the buds' potency, flavor, and yield. Use a magnifying tool to monitor the trichomes and determine when they reach peak maturity. Harvest when most trichomes have become milky white or cloudy, which indicates a high cannabinoid content. Some growers prefer to wait until some trichomes turn amber to achieve a more sedative effect.

2. Flushing:

Prior to harvesting, it's advisable to flush the plants using plain, pH-neutral water. This process eliminates surplus nutrients from both the growing medium and the plant tissues. Flushing contributes to the enhancement of the buds' flavor and scent by diminishing the residual nutrients.

3. Preparation:
Gather your harvesting equipment, such as sharp scissors or pruning shears, gloves, and sterile containers for holding the buds. Make sure that all tools and surfaces are clean and disinfected to avoid any contamination.

4. Selective Harvesting:
Selectively harvest mature buds instead of cutting down the whole plant simultaneously. Begin by picking the largest, ripest buds, allowing the smaller ones to grow and mature for a subsequent harvest.

5. Cutting:
Utilize sharp scissors or pruning shears to individually snip each bud from the plant. Ensure you make clean cuts near the main stem or branches to reduce harm to the plant and aid in the drying process.

6. Trimming:
Once the buds are cut, use scissors or trimming shears to remove any excess foliage and large fan leaves. Retain the smaller sugar leaves, as they are rich in trichomes and enhance the buds' potency.

7. Handling:
Carefully handle the harvested buds to prevent damage to the trichomes and the fragile structure of the flower. Refrain from squeezing or pressing the buds too hard to maintain their quality.

8. Drying:
Suspend the trimmed buds upside down in a cool, dark, and well-ventilated area to dry, maintaining a relative humidity of about 50-60%. Employ drying racks or lines to hang the buds, ensuring proper air flow around them.

9. Curing:
Once dried, move the buds into airtight glass jars to begin the curing process. Ensure the jars are sealed well and placed in a cool, dark

area with a relative humidity of approximately 60-65%. Periodically open the jars to let out any excess moisture and ensure effective curing.

10. Monitoring:
Closely monitor the drying and curing processes, regularly inspecting the buds for moisture content, mold, or mildew. Make necessary adjustments to the drying and curing conditions to achieve the best results.

Adhering to these harvesting methods allows hydroponic growers to enhance the quality, potency, and total yield of their buds. Accurate timing, meticulous preparation, precise cutting, careful trimming, and gentle handling are crucial to yield cannabis flowers of high quality that fulfill the expected standards.

Trimming and Manicuring Techniques

Trimming and manicuring buds is a crucial step in the harvesting process of hydroponic marijuana cultivation. Proper trimming ensures that the harvested buds are clean, visually appealing, and ready for drying and curing. Here's a guide to trimming and manicuring buds effectively:

1. Gather the Necessary Tools:
Before starting the trimming process, gather all the necessary tools and supplies. These may include sharp scissors or pruning shears, gloves, a trimming tray or table, and containers for collecting trimmed buds and excess foliage.

2. Prepare the Work Area:
Set up a clean and organized work area for trimming. Use a flat surface such as a table or trimming tray covered with a clean surface or parchment paper to catch trimmings and prevent mess.

3. Trimming Technique:
Grip the bud gently between your fingers and utilize sharp scissors or pruning shears to meticulously remove any excess foliage and

large fan leaves. Ensure you make clean cuts as near to the bud's base as possible, taking care not to harm the adjacent plant material. Discard any leaves lacking trichomes or resin glands, since they add harshness when smoked and diminish the buds' aesthetic appeal.

4.Manicuring Technique:
After removing the larger fan leaves, focus on manicuring the buds to remove any remaining sugar leaves and small stems. Sugar leaves are the smaller leaves that protrude from the buds and may contain trichomes.
Use precision trimming scissors or snips to carefully trim away excess foliage and sugar leaves, leaving behind only the densely packed buds with minimal leaf material.

5.Maintain Consistency:
Aim for consistency in the appearance and size of trimmed buds to ensure uniformity and quality in the final product. Trim buds to a similar size and shape, removing any irregularities or excess material for a clean and professional finish.

6.Handle Buds Gently:
Handle the trimmed buds with care to avoid damaging the delicate flower structure and trichomes. Avoid squeezing or compressing the buds excessively during handling to preserve their integrity and potency.

7.Inspect and Touch-Up:
After trimming, inspect each bud closely to ensure that it meets quality standards. Touch up any remaining leaf material or uneven edges with precision trimming scissors to achieve a polished appearance.

8.Collect Trimmed Buds:
As you trim each bud, collect the trimmed buds in clean containers or jars for storage. Avoid overcrowding or compressing the buds during storage to prevent damage and preserve their quality.

9.Dispose of Trimmed Material:
Dispose of the excess foliage and trimmings properly, either by composting or disposing of them in accordance with local

regulations. Keeping the work area clean and free of debris ensures a hygienic and efficient trimming process.

By following these trimming and manicuring techniques, hydroponic growers can produce clean, visually appealing buds with enhanced potency and flavor. Proper trimming is essential for maximizing the quality and marketability of the final product, whether for personal use or commercial sale.

7.2 CURING AND DRYING PROCESS

Importance of Curing

Curing is an essential step in the post-harvest process of hydroponic marijuana cultivation. It involves carefully drying and aging harvested buds in controlled conditions to enhance their flavor, potency, and overall quality. Curing plays a crucial role in unlocking the full potential of cannabis flowers and improving their shelf life. Here's why curing is important:

1. Enhanced Flavor and Aroma:

Curing facilitates the gradual decomposition of chlorophyll and other unwanted compounds in freshly harvested buds. This process softens the harsh, grassy flavor of new cannabis and improves its inherent taste and scent. As time passes, curing accentuates the distinct terpene profile of each variety, leading to a more intricate and pleasurable sensory experience. Buds that are properly cured tend to have deeper, more subtle flavors and scents than those that are not cured or are cured incorrectly.

2. Improved Potency:

During curing, cannabinoids such as THC and CBD undergo chemical changes that can increase their potency and bioavailability. Enzymes and beneficial bacteria present in the buds break down precursor molecules into more active forms, potentially boosting the overall psychoactive and therapeutic effects of the cannabis.

Properly cured buds are typically smoother to smoke or vaporize, with a more balanced and consistent potency throughout. Curing helps ensure that the cannabinoids are evenly distributed and fully activated, resulting in a more predictable and enjoyable user experience.

3. Reduced Harshness and Irritation:

Freshly harvested cannabis can be harsh and irritating to the throat and lungs due to the presence of residual moisture and chlorophyll. Curing allows excess moisture to evaporate gradually, resulting in a smoother, more palatable smoke or vapor.

Well-cured buds burn more evenly and cleanly, producing a smoother inhalation experience with less coughing or throat irritation. This makes cured cannabis more enjoyable and accessible to a wider range of consumers, including medical patients and recreational users.

4. Preservation of Freshness and Shelf Life:

Properly cured buds have a lower moisture content and are less susceptible to mold, mildew, and bacterial contamination. Curing helps stabilize the moisture content of the buds and creates an inhospitable environment for microbial growth, extending their shelf life and preserving freshness.

Well-cured buds can be stored for longer periods without degradation in quality, allowing growers to maintain a consistent supply of high-quality cannabis over time. This is particularly important for commercial cultivators who need to store and distribute their product effectively.

5. Enhanced Long-Term Storage:

Cured cannabis, when stored correctly, can retain its potency and quality for an extended time. Buds that are properly cured can preserve their strength, taste, and scent for several months or even years if kept in airtight containers and placed in a cool, dark environment.

Curing enables cannabis to experience a gradual, natural maturation process that enhances its quality and stability. As a result, cured cannabis becomes an excellent option for extended storage and maturation, offering aficionados the opportunity to savor the nuanced shifts in flavor and strength that emerge with extended curing.

In summary, the curing process is crucial in hydroponic cannabis cultivation, serving to improve the taste, strength, and overall caliber of the buds. Through careful drying and maturation of the cannabis flowers under regulated conditions, cultivators can realize the full capabilities of the plant, offering users a top-tier product that ensures an enhanced sensory encounter. Mastery of the curing process is indispensable for any grower aiming to yield cannabis of the highest quality, with optimal potency and consumer desirability.

Proper Drying Techniques

Proper drying techniques are essential in hydroponic marijuana cultivation to preserve the quality, potency, and overall integrity of harvested buds. Drying involves removing excess moisture from the buds while retaining their essential oils, cannabinoids, and terpenes. Here's a guide to proper drying techniques for hydroponic marijuana:

1. Harvest Timing:
Harvest the plants when they reach peak maturity, as determined by the color and clarity of the trichomes. Monitor trichomes using a magnifying tool and harvest when they are predominantly milky white or cloudy, with some amber-colored trichomes for desired effects.

2. Preparation:
After harvesting, carefully trim excess foliage and large fan leaves from the buds using sharp scissors or pruning shears. Leave smaller sugar leaves intact, as they contain trichomes and contribute to the potency of the buds.

3. Hang Drying:
Suspend the trimmed buds upside down in a cool, dark, and well-ventilated area for drying. Utilize racks or lines to hang the buds,

ensuring air can circulate freely around each one. Keep the relative humidity between 50-60% to inhibit the growth of mold and mildew.

4. Temperature Control:

Maintain the drying area at a moderate temperature, ideally between 60-70°F (15-21°C). High temperatures should be avoided because they may lead to rapid drying, which can diminish the flavor and potency of the buds. Similarly, low temperatures should be avoided as they can extend the drying time and heighten the risk of mold development.

5. Air Circulation:

To ensure proper air circulation in the drying area, utilize fans or natural ventilation. Adequate airflow is crucial for evenly removing moisture from the buds, which helps prevent mold and mildew. Arrange the fans to softly circulate the air around the buds, avoiding any excessive movement.

6. Check Moisture Levels:

Regular monitoring of the drying process is essential by checking the moisture content of the buds. They should feel dry on the outside but still contain some moisture inside. Avoid overdrying the buds, as this can result in a harsh smoke and reduced flavor. To check for moisture, gently squeeze a bud between your fingers; it should be springy and a bit tacky, indicating the right amount of moisture. If a bud is too dry or crumbles easily, it may have been overdried.

7. Patience:

It is advisable to let the buds dry gradually over a period of 7-14 days, taking into account the environmental conditions and the specific traits of the plant. Hastening the drying process can adversely affect the quality and strength of the buds. Patience is crucial for obtaining cannabis that is dried and cured correctly.

8. Bud Burping:

During the drying process, periodically "burping" the jars is crucial. This involves opening the jars daily for a few minutes to allow fresh air in and let excess moisture escape, which helps prevent mold formation. It's essential to monitor the buds closely for any signs of mold or mildew and act swiftly if any problems are detected.

9.Storage:

After the buds are adequately dried, place them in airtight glass jars to begin the curing process. Keep the jars in a cool, dark area with a relative humidity of approximately 60-65%. Regularly check the buds throughout the curing period and modify the conditions if necessary to achieve the best outcome.

Adhering to the correct drying techniques, hydroponic cultivators can maintain the quality, potency, and integrity of their harvested buds. Proper drying is crucial for yielding cannabis flowers of high quality, with improved flavor, aroma, and effects.

Curing Jars and Storage

Curing jars and storage containers are essential components of the curing process in hydroponic marijuana cultivation. Properly selected and maintained jars ensure optimal conditions for curing, preserving the flavor, potency, and overall quality of the harvested buds. Here's what you need to know about curing jars and storage containers:

1.Glass Jars:

Glass jars are the preferred choice for curing cannabis buds due to their non-reactive nature and ability to maintain a stable environment.

Mason jars or canning jars with airtight seals are commonly used for curing. Choose jars with wide mouths for easy access and handling of the buds.

2.Size:

Select curing jars that match the quantity of buds harvested, ensuring they are not overly packed to allow adequate airflow and prevent moisture buildup. It is better to use multiple smaller jars rather than one large jar to reduce the risk of mold and mildew in case of issues with certain batches.

3. Airtight Seals:
Ensure that the curing jars have airtight seals to prevent any exchange of air and moisture with the external environment. Proper sealing is crucial for maintaining consistent humidity levels inside the jars, which is imperative for the curing process. Inspect the seals regularly for any damage or wear and replace them immediately to maintain optimal curing conditions.

4. Opaque or Dark-Colored Jars:
Consider using opaque or dark-colored glass jars to protect the buds from exposure to light during curing. Light exposure can degrade cannabinoids and terpenes, compromising the quality and potency of the buds.
Alternatively, store clear glass jars in a dark place or cover them with a cloth or opaque material to block out light and prevent degradation of the buds.

5. Humidity Packs:
Some growers choose to include humidity packs or moisture-control packets in the curing jars to help regulate humidity levels and prevent overdrying or mold growth.
Humidity packs are available in various humidity levels (e.g., 58%, 62%, 65%) and can help maintain optimal moisture content in the buds during curing.

6. Labeling:
Ensure each curing jar is labeled with the strain name, harvest date, and any pertinent details to track various batches and manage rotation effectively. Accurate labeling facilitates the identification of each jar's contents and simplifies the monitoring of the curing process and the evaluation of quality over time.

7. Storage Conditions:
Store curing jars in a cool, dark location with consistent temperature and humidity. Avoid direct sunlight, heat sources, or temperature changes, as these can impact the curing process and the quality of

the buds. Keep the jars away from strong odors to prevent the buds from absorbing undesirable flavors or smells.

Using the right curing jars and storage methods, hydroponic growers can ensure optimal conditions for curing cannabis buds, which enhances flavor, potency, and overall quality. Buds that are properly cured and stored in sealed glass jars can retain their freshness and therapeutic qualities for a longer time, providing growers with a superior product for maximum enjoyment and benefit.

CHAPTER 8

TROUBLESHOOTING COMMON ISSUES

8.1 NUTRIENT DEFICIENCIES AND EXCESSES

Identifying Nutrient Imbalances

Identifying nutrient imbalances in hydroponic marijuana cultivation is crucial for maintaining plant health, optimizing growth, and maximizing yields. Nutrient imbalances can manifest in various ways, affecting plant appearance, growth patterns, and overall vigor. Here are some common signs to look for when identifying nutrient imbalances:

1. Leaf Discoloration:
Yellowing or discoloration of leaves is a common indicator of nutrient imbalances. Different nutrient deficiencies or excesses can cause specific patterns of discoloration.
Nitrogen deficiency: Older leaves turn yellow starting from the tips and edges, progressing towards the veins, while the leaf veins remain green.

Phosphorus deficiency: Leaves may develop dark green or purple discoloration, often starting from the tips and edges and spreading inward.

Potassium deficiency: Leaf margins may become scorched or brown, and older leaves may exhibit yellowing or necrosis starting from the leaf edges.

Calcium deficiency: New leaves may appear distorted or deformed, and leaf tips may die back, exhibiting a burnt appearance.

Magnesium deficiency: Interveinal chlorosis develops on older leaves, with yellowing between the leaf veins while the veins remain green.

2. Stunted Growth or Poor Development:
Nutrient imbalances can inhibit plant growth and development, resulting in stunted growth, reduced vigor, and poor overall plant health.

Insufficient nutrients can lead to slow or stunted growth, smaller leaves, and reduced internodal spacing, while excessive nutrients may cause leaf curling, brittle stems, and abnormal growth patterns.

3. Leaf Symptoms:
Leaf symptoms such as curling, cupping, or twisting can indicate nutrient imbalances. For example, calcium deficiency may cause leaf cupping or distortion, while magnesium deficiency can result in leaf curling or twisting.

4. Bud Development:
Nutrient imbalances can affect bud development and flower formation, leading to reduced yields, smaller buds, or poor bud structure.

Insufficient phosphorus or potassium may result in reduced bud size and density, while excess nitrogen can delay flowering and reduce bud quality.

5. Root Symptoms:
Monitoring root health and development can also provide clues about nutrient imbalances. For example, root rot or decay may

indicate overwatering or oxygen deficiency, while healthy white roots suggest adequate nutrient uptake.

Brown or discolored roots may indicate root rot or nutrient imbalances, while healthy white roots suggest optimal nutrient conditions.

6. Overall Plant Appearance:

Pay attention to the overall appearance of the plants, including leaf size, color, and shape, as well as stem thickness, internodal spacing, and overall plant vigor.

Healthy plants exhibit lush green foliage, vigorous growth, and robust stems, while plants with nutrient imbalances may appear stunted, discolored, or unhealthy.

7. pH and EC Levels:

Monitoring pH and electrical conductivity (EC) levels of the nutrient solution can help identify nutrient imbalances and maintain optimal nutrient uptake by the plants.

pH levels outside the recommended range (typically 5.5-6.5 for hydroponic systems) can affect nutrient availability and uptake, leading to nutrient imbalances.

EC levels that are too high or too low may indicate nutrient excesses or deficiencies, respectively, affecting plant growth and health.

By observing these signs and symptoms, hydroponic growers can identify and address nutrient imbalances promptly, ensuring optimal plant health, growth, and yield. Regular monitoring of nutrient levels, pH, and plant health is essential for maintaining a healthy hydroponic system and maximizing the success of marijuana cultivation.

Corrective Measures

Correcting nutrient imbalances in hydroponic marijuana cultivation is essential for maintaining plant health, promoting optimal growth, and maximizing yields. Once nutrient imbalances are identified through careful observation and monitoring, corrective measures can be implemented to restore balance and support healthy plant development. Here are some effective corrective measures for addressing nutrient imbalances in hydroponic systems:

1. Adjusting Nutrient Solution:
To correct nutrient imbalances, measure and adjust the nutrient solution. Utilize pH meters and EC meters to monitor and regulate the pH and nutrient levels. Incorporate suitable hydroponic nutrients or supplements into the reservoir, ensuring plants receive the necessary nutrients in accurate proportions. Adhere to the manufacturer's recommendations and nutrient guidelines to maintain proper nutrient balance and prevent over or under-fertilization.

2. pH Adjustment:
Monitoring and adjusting the pH level of the nutrient solution is vital for optimizing nutrient availability and plant absorption. Use pH-up or pH-down solutions to adjust the pH as needed. Maintaining the pH of the nutrient solution within the optimal range for hydroponic growth, typically between 5.5 and 6.5 for most plants, is crucial because pH levels outside this range can affect nutrient availability and lead to nutrient imbalances.

3. Flush the System:
If nutrient imbalances are severe or persistent, consider flushing the hydroponic system with plain, pH-balanced water to remove excess salts and reset nutrient levels.
Flushing helps remove built-up nutrient residues and allows for a fresh start, promoting better nutrient uptake and plant health.

4. Optimize Water Quality:
Ensure that the water source used for preparing the nutrient solution is clean, free from contaminants, and has suitable mineral content for hydroponic cultivation.
Use filtered or purified water to minimize the risk of nutrient imbalances caused by impurities or excessive mineral content in the water.

5. Monitor Environmental Conditions:

Ensuring optimal environmental factors, including ventilation, humidity, and temperature, in a grow room or greenhouse is crucial for promoting robust plant development and efficient uptake of nutrients. Heat stress and humidity can impede nutrient absorption, leading to issues with air circulation. Therefore, ensuring adequate ventilation and addressing nutritional imbalances are essential in preventing these complications.

6. Inspect and Adjust Lighting:

Evaluate the lighting setup and ensure that plants receive adequate light intensity and duration for optimal photosynthesis and nutrient uptake.

Adjust the distance, intensity, and photo period of grow lights as needed to prevent light-related stress and promote balanced nutrient uptake.

7. Address Root Zone Issues:

Inspect the root zone for signs of root rot, decay, or other issues that may affect nutrient absorption. Maintain proper aeration, oxygenation, and moisture levels in the root zone to support healthy root development and nutrient uptake.

Consider using beneficial microbial inoculants or root supplements to promote a healthy rhizosphere and improve nutrient availability to the plants.

8. Monitor Plant Health:

Continuously monitor plant health, growth patterns, and nutrient status to detect and address nutrient imbalances early.

Regularly inspect leaves, stems, and roots for signs of nutrient deficiencies, excesses, or other nutrient-related issues, and take corrective actions promptly to prevent further damage.

By implementing these corrective measures, hydroponic growers can effectively address nutrient imbalances and maintain optimal nutrient levels for healthy, vigorous plant growth. Regular monitoring, adjustment, and maintenance of nutrient levels and environmental conditions are essential for preventing nutrient imbalances and promoting successful hydroponic cultivation of marijuana.

8.2 PEST AND DISEASE MANAGEMENT

Recognizing Symptoms

Recognizing symptoms of nutrient deficiencies or excesses in hydroponic marijuana cultivation is crucial for maintaining plant health and maximizing yields. By closely monitoring plant growth and appearance, growers can identify early signs of nutrient imbalances and take corrective actions promptly. Here are some common symptoms to look for when recognizing nutrient deficiencies or excesses:

1. Nitrogen (N) Deficiency:
Symptoms: Yellowing (chlorosis) of lower leaves starting from the tips and edges, gradual progression towards the veins, leaves may become pale or light green overall. Stunted growth and reduced vigor.
Cause: Insufficient nitrogen in the nutrient solution.
Corrective Action: Increase nitrogen levels in the nutrient solution or use nitrogen-rich fertilizers.

2. Phosphorus (P) Deficiency:
Symptoms: Dark green or purplish discoloration of leaves, especially on older leaves. Leaves may develop a bronze or reddish tint. Reduced growth and development, delayed flowering.
Cause: Insufficient phosphorus in the nutrient solution.
Corrective Action: Increase phosphorus levels in the nutrient solution or use phosphorus-rich fertilizers.

3. Potassium (K) Deficiency:

Symptoms: Marginal scorching or browning of leaf edges, yellowing or necrosis of older leaves, weak stems, and poor bud development.
Cause: Insufficient potassium in the nutrient solution.
Corrective Action: Increase potassium levels in the nutrient solution or use potassium-rich fertilizers.

4.Calcium (Ca) Deficiency:
Symptoms: Distorted or deformed new growth, curling or cupping of leaves, brown spots or necrosis on leaf margins, weak stems.
Cause: Insufficient calcium in the nutrient solution.
Corrective Action: Increase calcium levels in the nutrient solution or use calcium-rich supplements.

5.Magnesium (Mg) Deficiency:
Symptoms: Interveinal chlorosis on older leaves, yellowing between leaf veins while the veins remain green, leaf curling or twisting.
Cause: Insufficient magnesium in the nutrient solution.
Corrective Action: Increase magnesium levels in the nutrient solution or use magnesium-rich supplements.

6.Iron (Fe) Deficiency:
Symptoms: Yellowing between leaf veins, known as interveinal chlorosis, often starting with new growth. Leaves may develop a whitish or pale appearance.
Cause: Insufficient iron in the nutrient solution or poor iron availability due to high pH.
Corrective Action: Increase iron levels in the nutrient solution or use chelated iron supplements.

7.Excess Nutrients:
Symptoms: Leaf tip burn or necrosis, leaf margins may become scorched or brown, leaf curling or cupping, stunted growth.
Cause: Over fertilization or excessive nutrient buildup in the growing medium.
Corrective Action: Flush the growing medium with plain water to remove excess salts, adjust nutrient levels in the solution, and avoid overfeeding.

8. pH Imbalance:
Symptoms: Nutrient deficiencies or excesses may be exacerbated by pH imbalances. pH levels outside the optimal range (typically 5.5-6.5 for hydroponic systems) can affect nutrient availability and uptake.
Corrective Action: Adjust pH levels in the nutrient solution using pH-up or pH-down solutions to bring them within the recommended range.

By recognizing these symptoms and understanding their underlying causes, hydroponic growers can take appropriate corrective actions to address nutrient imbalances and maintain optimal plant health and productivity. Regular monitoring, adjustment of nutrient levels, and proper management of environmental conditions are essential for preventing and correcting nutrient-related issues in hydroponic marijuana cultivation.

Treatment Options

Treatment options for nutrient deficiencies or excesses in hydroponic marijuana cultivation depend on the specific nutrient imbalance identified. Here are some general treatment options for addressing nutrient issues:

1. Adjust Nutrient Solution:
Increase or decrease the concentration of specific nutrients in the nutrient solution to correct deficiencies or excesses. Follow manufacturer's guidelines and nutrient recommendations to ensure proper dosing and application.
Use balanced hydroponic nutrient formulations or supplements designed to address specific nutrient deficiencies while maintaining overall nutrient balance.

2. pH Adjustment:
To optimize nutrient availability and absorption for the plants, adjust the pH of the nutrient solution. Utilize pH-up or pH-down solutions to increase or decrease the pH accordingly, ensuring it falls within

the ideal range for hydroponic cultivation, which is typically between 5.5 and 6.5.

Consistently monitoring and maintaining the pH levels in your nutrient solution is crucial to prevent nutrient lockout and imbalances that can occur due to pH fluctuations.

3. Flush System:

Flush the hydroponic system with plain, pH-balanced water to remove excess salts, nutrients, or impurities from the growing medium and root zone. Flushing helps reset nutrient levels and prevent nutrient buildup or imbalances.

Monitor the runoff water to gauge the effectiveness of flushing and ensure that excess nutrients are adequately removed from the system.

4. Adjust Environmental Conditions:

For maximum plant development and effective nutrient absorption, environmental parameters like temperature, humidity, and air circulation must be balanced. In order to avoid heat stress and humidity-related issues that might hinder the absorption of nutrients, proper ventilation and air circulation are essential. It is crucial to make sure that the length and intensity of the light meet the needs of photosynthesis and nutritional absorption. It is critical to modify lighting conditions as needed to protect plants from light stress and to promote their health.

5. Supplemental Feeding:

Provide supplemental feeding with foliar sprays or root drenches containing specific nutrients to address deficiencies or boost nutrient levels. Choose nutrient-rich supplements or fertilizers formulated for hydroponic cultivation to ensure compatibility with the growing system.

Apply supplements according to recommended dosage rates and application methods to avoid overfeeding or nutrient toxicity.

6. Root Zone Treatments:

Treat the root zone with beneficial microbial inoculants or root supplements to promote healthy root development and improve nutrient uptake. Beneficial microbes can enhance nutrient availability and uptake by facilitating nutrient cycling and decomposition of organic matter.

Apply root treatments as directed by the manufacturer to ensure proper colonization and establishment of beneficial microorganisms in the root zone.

7. Monitor and Adjust:

Continuously monitor plant health, growth patterns, and nutrient levels to assess the effectiveness of treatment options and make necessary adjustments.

Regularly inspect leaves, stems, and roots for signs of nutrient deficiencies, excesses, or other nutrient-related issues, and take corrective actions promptly to prevent further damage.

By implementing these treatment options and addressing nutrient imbalances promptly, hydroponic growers can maintain optimal nutrient levels, support healthy plant growth, and maximize yields in their marijuana cultivation. Regular monitoring, adjustment of nutrient levels, and proper management of environmental conditions are essential for preventing and correcting nutrient-related issues in hydroponic systems.

CHAPTER 9

ADVANCED TECHNIQUES AND INNOVATIONS

9.1 HYDROPONICS SYSTEMS INNOVATIONS

Aeroponics

Aeroponics is a cutting-edge hydroponic farming method that grows plants in an environment rich in air, with roots hanging in the air and frequently sprayed with a nutrient-dense solution. In contrast to conventional hydroponic systems that submerge plants in a nutrient solution, aeroponics administers nutrients straight to the roots via a fine mist, facilitating maximum nutrient uptake and aeration. Below is a summary of aeroponics and its principal characteristics:

1. Root Zone Environment:
In aeroponics, plant roots are suspended in a chamber or container, typically made of plastic or another inert material, where they are exposed to air. This air-rich environment promotes rapid root growth and allows for efficient nutrient absorption.

2. Nutrient Delivery:
Nutrients are delivered to the roots as a fine mist or aerosol spray using a high-pressure pump and misting nozzles. The nutrient

solution is atomized into tiny droplets, which are then sprayed directly onto the roots at regular intervals.

The misting schedule can be controlled using timers or automated systems to ensure consistent nutrient delivery and optimal root hydration.

3. Oxygenation:

Aeroponic systems provide ample oxygen to the roots, promoting aerobic respiration and healthy root development. The misting process creates a highly oxygenated environment around the roots, allowing for efficient nutrient uptake and minimizing the risk of anaerobic conditions.

4. Water Efficiency:

Aeroponic systems are highly water-efficient compared to traditional soil-based or hydroponic cultivation methods. Because the roots are suspended in the air and misted with a nutrient solution, there is minimal water wastage through evaporation or runoff.

The closed-loop design of aeroponic systems allows for precise control over water usage, making them ideal for arid or water-scarce environments.

5. Space Efficiency:

Aeroponic systems are space-efficient and can be configured to fit a wide range of indoor or outdoor growing spaces. The vertical design of aeroponic towers or racks maximizes space utilization, allowing for high plant density and increased crop yields per square foot.

6. Disease Prevention:

Aeroponic systems minimize the risk of soil-borne diseases and pathogens commonly associated with traditional soil-based cultivation. The absence of soil eliminates the need for soil sterilization and reduces the likelihood of root rot, fungal infections, and other soil-borne ailments.

Additionally, the high oxygen levels in aeroponic systems inhibit the growth of anaerobic pathogens, further enhancing plant health and disease resistance.

7. Plant Growth and Yields:

Aeroponics accelerates plant growth and development, resulting in higher yields and reduced crop cycles. The effective supply of nutrients and oxygen to the roots encourages robust vegetative growth, blossoming, and fruit production, leading to healthier plants and superior harvests.

8. Automation and Control:
Aeroponic systems can achieve full automation and control through sophisticated monitoring and management technologies. With automated nutrient dosing systems, pH controllers, and environmental sensors, it is possible to maintain precise control over the growing conditions, thereby optimizing plant growth and productivity.

Aeroponics offers several advantages over traditional cultivation methods, including water efficiency, space efficiency, disease prevention, and accelerated plant growth. By harnessing the power of air and mist to deliver nutrients directly to the roots, aeroponic systems provide an innovative solution for sustainable and high-yield crop production in a variety of environments.

Deep Water Culture

Deep Water Culture (DWC) is a widely-used hydroponic farming method where plant roots are suspended in a nutrient-dense solution, optimizing nutrient absorption and oxygenation. DWC setups immerse the roots in a deep, oxygen-infused nutrient reservoir, which encourages swift growth and robust plant health. Below is a summary of DWC and its principal characteristics:

1. Reservoir Setup:
DWC systems consist of a reservoir or container filled with a nutrient solution. The reservoir is typically made of plastic or another inert material and is deep enough to submerge the plant roots fully.

An air pump and air stone are used to aerate the nutrient solution, ensuring adequate oxygenation for the roots and preventing anaerobic conditions.

2. Plant Support:

Plants are supported above the reservoir using a floating platform or net pot suspended in the nutrient solution. The plant's roots extend into the solution below, where they absorb water and nutrients directly.

3. Nutrient Solution:

The nutrient solution in DWC systems is carefully balanced to provide essential macro and micronutrients required for plant growth. Nutrient concentrations are monitored and adjusted regularly to maintain optimal levels for healthy plant development.

pH levels of the nutrient solution are also monitored and adjusted to ensure proper nutrient uptake and prevent nutrient deficiencies or toxicities.

4. Oxygenation:

Aeration is essential in DWC systems to maintain adequate oxygen levels in the nutrient solution. An air pump is used to deliver oxygen to the roots through air stones or diffusers placed in the reservoir. Oxygenation of the nutrient solution promotes healthy root development and prevents root suffocation or rot caused by oxygen deprivation.

5. Water Efficiency:

DWC systems are water-efficient compared to traditional soil-based cultivation methods since water is recirculated within the system and not lost through evaporation or runoff.

The closed-loop design of DWC systems minimizes water wastage and allows for precise control over water usage, making them ideal for water-conscious growers.

6. Space Efficiency:

Deep Water Culture (DWC) systems are designed for space efficiency and can be installed in diverse indoor or outdoor growing settings. Their compact structure enables a higher density of plants, leading to greater crop yields per square foot.

7. Plant Growth and Yields:

Deep Water Culture (DWC) enhances swift plant growth and development, resulting in higher yields and reduced crop cycles. The immediate availability of water and nutrients enables vigorous plant growth and the production of abundant, robust harvests. DWC systems are especially advantageous for rapidly growing plants like leafy greens, herbs, and some flowering crops, including cannabis.

8. Automation and Control:

DWC systems can be automated and controlled using timers, sensors, and nutrient dosing systems. Automated pH controllers, nutrient pumps, and environmental monitors allow for precise control over growing conditions, optimizing plant growth and productivity.

Overall, Deep Water Culture (DWC) is a highly effective hydroponic cultivation technique that offers several advantages, including water efficiency, space efficiency, and accelerated plant growth. By providing plants with direct access to water, nutrients, and oxygen, DWC systems support healthy root development and robust plant growth, making them a popular choice for hydroponic growers seeking high yields and quality harvests.

Nutrient Film Technique (NFT)

By continuously exposing plant roots to a nutrient-rich fluid, the Nutrient Film Technique (NFT) is a hydroponic method that enhances nutrient absorption and oxygenation. Plants are placed in troughs or tubes with a gentle slope, allowing a thin layer of nutritional solution to cover the roots before being recycled back into the reservoir. An outline of NFT and its main characteristics is provided below:

1. Channel Setup:

NFT systems consist of channels or gullies made of inert materials such as PVC, plastic, or metal. These channels are typically arranged in a horizontal or slightly sloped configuration to allow for gravity-driven flow of the nutrient solution.

Plants are placed in small net pots or grow cups filled with an inert growing medium such as rockwool or perlite, which are then inserted into holes or slots in the channels. The roots of the plants extend into the channels, where they are continuously bathed in a thin film of nutrient solution.

2.Continuous Flow:

A pump circulates the nutrient solution from the reservoir through tubing or pipes to the top end of the channels. The solution flows down the channels, nourishing the plant roots before returning to the reservoir. This nutrient solution circulates continuously in a closed-loop system, ensuring a steady supply of water, nutrients, and oxygen to the plants.

3.Nutrient Solution:

In NFT systems, the nutrient solution is meticulously balanced to supply the necessary macro and micro-nutrients for plant growth. The concentrations of nutrients are consistently monitored and fine-tuned to sustain optimal conditions for robust plant development. Similarly, the pH levels of the nutrient solution are regularly checked and modified to facilitate proper nutrient absorption and avert any deficiencies or toxicities.

4.Oxygenation:

In NFT systems, oxygenating the nutrient solution is vital to prevent root suffocation and encourage robust root growth. Aeration tools, like air stones or diffusers, are commonly utilized to infuse the nutrient solution with oxygen in the reservoir. The slender layer of nutrient solution that flows across the roots delivers sufficient oxygen to the root area, facilitating effective nutrient absorption and vigorous plant development.

5.Water Efficiency:

NFT systems are more water-efficient than traditional soil-based cultivation methods because the water is recirculated within the system, preventing loss through evaporation or runoff. The closed-loop design of NFT systems reduces water waste and enables precise control over water use, making them suitable for growers concerned about water conservation.

6. Space Efficiency:
Nutrient Film Technique (NFT) systems are designed for space efficiency and can be installed in diverse indoor or greenhouse settings. Their streamlined configuration of NFT channels facilitates a higher density of plants, leading to greater yields per square foot.

7. Plant Growth and Yields:
NFT enhances quick plant growth and development, resulting in higher yields and reduced crop cycles. The uninterrupted provision of water, nutrients, and oxygen to the roots enables plants to thrive and yield abundant, robust harvests. NFT systems are especially advantageous for rapidly maturing crops such as leafy greens, herbs, and specific flowering plants, like strawberries and lettuce.

8. Automation and Control:
NFT systems can be automated and controlled using timers, sensors, and nutrient dosing systems. Automated pH controllers, nutrient pumps, and environmental monitors allow for precise control over growing conditions, optimizing plant growth and productivity.

Overall, Nutrient Film Technique (NFT) is a highly effective hydroponic cultivation method that offers several advantages, including water efficiency, space efficiency, and accelerated plant growth. By providing plants with a constant flow of water, nutrients, and oxygen, NFT systems support healthy root development and robust plant growth, making them a popular choice for hydroponic growers seeking high yields and quality harvests.

9.2 HIGH-TECH GROWING EQUIPMENT

Automated Systems

Automated systems are vital in contemporary hydroponic farming, providing cultivators with the capability to precisely and efficiently oversee and adjust different elements of the growth environment. These systems employ sophisticated technologies and sensors to manage tasks like nutrient delivery, pH balance, temperature and humidity levels, light cycles, and watering automatically. Below is a summary of the automated systems employed in hydroponic agriculture:

1. Nutrient Dosing Systems:
Automated nutrient dosing systems accurately measure and deliver the required amounts of nutrients to the nutrient solution based on predefined parameters and setpoints.
These systems typically consist of peristaltic pumps or dosing pumps controlled by a central controller or computerized system. Nutrient solutions are mixed in precise ratios and delivered to the hydroponic reservoir or irrigation system as needed.

2. pH Regulation Systems:
Self-regulating devices are designed to monitor and adjust the pH levels of nutrient solutions, ensuring they remain within the ideal range for plant uptake. pH controllers utilize dosing pumps and sensors to continuously measure the solution's pH, dispensing or withholding pH modifiers as needed to maintain the target range.

3. Environmental Monitoring:
Automated environmental monitoring systems measure and record key parameters such as temperature, humidity, CO_2 levels, and light intensity in the growing environment.
Sensors placed throughout the grow room or greenhouse collect real-time data, which is then analyzed and used to adjust environmental conditions as necessary to optimize plant growth and health.

4. Climate Control Systems:

Climate control systems are designed to regulate environmental factors like temperature and humidity, thereby creating ideal conditions for plant growth.

Automated HVAC systems, along with evaporative coolers, dehumidifiers, and humidifiers, utilize sensor data to regulate and maintain stable and optimal conditions for growth.

5. Irrigation and Watering Systems:

Automated irrigation and watering systems deliver water and nutrient solution to plants at scheduled intervals or based on soil moisture levels.

Drip irrigation, micro-sprinklers, or misting systems are controlled by timers or sensor-based controllers to ensure uniform watering and efficient nutrient delivery to plant roots.

6. Lighting Control Systems:

Lighting control systems regulate the timing, intensity, and spectrum of artificial lighting in indoor growing environments.

Automated timers or programmable controllers adjust the duration and intensity of grow lights to mimic natural sunlight and optimize photosynthesis during different growth stages.

7. Data Logging and Analysis:

Automated systems often include data logging and analysis capabilities to record and analyze environmental data over time.

Growers can access historical data and trends to track plant growth, monitor system performance, and make informed decisions about adjustments and optimizations.

8. Remote Monitoring and Control:

Some automated systems offer remote monitoring and control capabilities, allowing growers to access and adjust settings from anywhere with an internet connection.

Mobile apps or web-based interfaces provide real-time updates and alerts, enabling growers to manage their hydroponic systems conveniently and efficiently.

Automated systems offer numerous benefits to hydroponic growers, including increased efficiency, precise control over growing conditions, and reduced labor requirements. By automating repetitive tasks and optimizing environmental parameters, these systems help maximize yields, improve crop quality, and streamline the cultivation process.

Environmental Sensors

Environmental sensors play a critical role in hydroponic marijuana cultivation, providing real-time data on key environmental parameters to help growers monitor and optimize growing conditions. These sensors measure various factors such as temperature, humidity, CO_2 levels, light intensity, and nutrient concentrations, allowing growers to maintain optimal conditions for plant growth and health. Here's how environmental sensors are used in hydroponic marijuana cultivation:

1.Temperature Sensors:
Temperature sensors monitor the air and nutrient solution temperature in the grow room or greenhouse.
Optimal temperature ranges promote healthy plant growth and metabolism, while extremes can stress plants and affect yields.
Data from temperature sensors help growers adjust heating, cooling, and ventilation systems to maintain stable temperature levels.

2.Humidity Sensors:
Humidity sensors measure the moisture content of the air in the growing environment.
Proper humidity levels are essential for preventing issues such as mold, mildew, and pest infestations.
Humidity sensors help growers adjust ventilation, dehumidification, or humidification systems to maintain optimal humidity levels for plant growth.

3.CO_2 Sensors:
CO_2 sensors track the levels of carbon dioxide in a grow room or greenhouse. This gas is crucial for photosynthesis, and higher CO_2

levels can boost plant growth and yield. These sensors assist growers in fine-tuning their CO2 supplementation tactics to guarantee that plants get sufficient carbon dioxide throughout their growth cycle.

4. Light Sensors:

Light sensors are used to gauge the intensity and duration of light within a growth environment. Adequate lighting is vital for photosynthesis and the development of plants, with each growth stage demanding specific light intensities. These sensors enable cultivators to fine-tune their artificial lighting to ensure the appropriate spectrum, intensity, and duration of light for the best plant growth.

5. pH and EC Sensors:

pH and EC (electrical conductivity) sensors monitor the pH level and nutrient concentration of the hydroponic nutrient solution.

Proper pH and nutrient levels are essential for nutrient uptake and plant health.

pH and EC sensors help growers maintain stable nutrient solution parameters by alerting them to any fluctuations or imbalances that may occur.

6. Water Level Sensors:

Water level sensors monitor the level of the nutrient solution in hydroponic reservoirs or irrigation systems.

Proper water levels ensure continuous nutrient delivery to plant roots and prevent dry or flooded conditions.

Water level sensors help growers maintain optimal nutrient solution levels and prevent system failures due to water shortages or overflows.

7. Data Logging and Analysis:

Environmental sensors frequently come equipped with data logging features to chronicle and preserve environmental information over time. Data analysis instruments enable cultivators to scrutinize trends, discern patterns, and make educated choices regarding the modification of growing conditions to enhance plant development and yields.

The use of environmental sensors in hydroponic cannabis cultivation allows cultivators to exert exact control over the growing environment, reduce plant stress, and increase yield. Real-time surveillance and data examination provide cultivators with the ability to implement prompt alterations and improvements, leading to more robust plants and superior harvest quality.

Data Driving Cultivation

Data-driven cultivation in hydroponic marijuana production involves using collected data from various sensors and monitoring systems to make informed decisions and optimize growing conditions for maximum yield and quality. Here's how data-driven cultivation works:

1. Data Collection:
Environmental sensors, such as those for temperature, humidity, CO_2 levels, light intensity, pH, EC, and water levels, continuously collect data from the growing environment and hydroponic system. These sensors transmit real-time data to a central monitoring system or software platform for analysis and interpretation.

2. Data Analysis:
Growers analyze the collected data to identify trends, patterns, and correlations between environmental parameters and plant performance.
Data analysis tools and algorithms help growers extract actionable insights and make informed decisions about adjustments and optimizations.

3. Optimization Strategies:
Based on data analysis, growers implement optimization strategies to fine-tune growing conditions and maximize plant growth and yields. Optimization strategies may include adjusting environmental parameters (temperature, humidity, CO_2 levels, light intensity),

nutrient concentrations, pH levels, irrigation schedules, and nutrient dosing rates.

4. Dynamic Control:
Growers use automated systems and controls to dynamically adjust growing conditions in response to real-time data and analysis. Automated systems regulate environmental parameters, nutrient delivery, irrigation, lighting schedules, and other factors to maintain optimal conditions for plant growth and development.

5. Iterative Process:
Data-driven cultivation is an iterative process, where growers continuously collect, analyze, and act on data to improve cultivation practices over time.
Growers monitor plant responses to adjustments and optimizations, refine strategies based on observed outcomes, and iterate the process to achieve optimal results.

6. Predictive Analytic:
Advanced data analytics and modeling techniques enable growers to predict plant behavior and anticipate future growing conditions. Predictive analytics help growers proactively address potential issues, optimize resource allocation, and forecast yields and harvest timing with greater accuracy.

7. Remote Monitoring and Management:
Growers can remotely monitor and manage their cultivation operations through mobile apps or web-based platforms. This remote access enables them to stay informed about their facilities, receive instant alerts and notifications, and make necessary adjustments promptly, no matter where they are.

8. Continuous Improvement:
Data-driven cultivation promotes ongoing enhancement and innovation in hydroponic cannabis production. By utilizing data insights, growers can optimize resource use, cut expenses, mitigate risks, and improve both the yield and quality of the crop.
Utilizing data-driven cultivation techniques, hydroponic cannabis growers can fine-tune their growing environments, enhance yields,

and consistently produce superior crops. These methods enable cultivators to make decisions grounded in evidence, adjust to evolving conditions, and realize increased efficiency and success in their growing operations.

CHAPTER 10

LEGAL CONSIDERATIONS AND COMPLIANCE

10.1 UNDERSTANDING MARIJUANA LAWS

Federal VS State Regulations

Federal and state regulations on marijuana cultivation, distribution, and usage vary significantly, creating a complex legal landscape for growers. Here is a summary of the differences between federal and state regulations:

1. Federal Regulations:

Marijuana remains illegal at the federal level in the United States under the Controlled Substances Act (CSA) as a Schedule I controlled substance.

The federal government classifies marijuana as a drug with a high potential for abuse and no accepted medical use, making it illegal to manufacture, distribute, or possess.

The Drug Enforcement Administration (DEA) enforces federal marijuana laws and regulations, and individuals or businesses involved in marijuana cultivation or distribution can face federal prosecution, fines, and imprisonment.

Federal regulations apply across the entire country and supersede state laws, meaning that federal authorities can enforce marijuana laws in states where marijuana is legal under state law.

2. State Regulations:

Many states in the US have passed laws legalizing medical and/or recreational marijuana use despite federal prohibition.
State regulations vary widely, with some states allowing both medical and recreational marijuana, some allowing only medical use, and others prohibiting marijuana entirely.
State regulations govern licensing, cultivation, distribution, possession limits, taxation, and other aspects of the marijuana industry within state borders.
State regulatory agencies, such as state departments of health or cannabis regulatory commissions, oversee the implementation and enforcement of state marijuana laws.
States with legal marijuana markets have created comprehensive regulatory frameworks to ensure public safety, product quality, and compliance with state laws.

3. Interplay Between Federal and State Laws:
The conflict between federal and state marijuana laws creates legal uncertainties and challenges for growers, businesses, and consumers operating in the cannabis industry.
While federal authorities have the authority to enforce federal marijuana laws, they have generally deferred enforcement to states that have legalized marijuana, particularly in states with robust regulatory systems.
However, federal enforcement actions can still occur, especially in cases involving interstate trafficking, organized crime, or violations of federal priorities such as sales to minors or diversion to states where marijuana remains illegal.
The federal government has also taken steps to address conflicts between federal and state marijuana laws, such as issuing guidance memos and allowing states to implement their own regulatory systems without federal interference.
In summary, federal regulations classify marijuana as illegal, while many states have legalized marijuana for medical and/or recreational use under their own laws. The interplay between federal and state

regulations creates a complex legal environment for marijuana growers, requiring careful compliance with both federal and state laws to operate legally and minimize legal risks.

Licensing and Permit Requirements

Licensing and permit requirements for marijuana cultivation vary depending on the jurisdiction and the type of cultivation (medical or recreational). Here's an overview of the general process and considerations for obtaining licenses and permits for marijuana cultivation:

1.Research Local Regulations:
Prior to initiating a marijuana cultivation business, it is crucial to investigate and comprehend the distinct licensing and permit stipulations within your locality. Regulatory conditions can differ markedly across states, counties, and cities.

2.Determine License Type:
Identify the appropriate cultivation license type for your operation by considering the intended use of the marijuana, whether for medical or recreational purposes, and the scale of the operation, such as cultivation, processing, or distribution. Each license type may vary in its requirements and the process for application.

3.Meet Eligibility Criteria:
Make sure to fulfill the eligibility criteria established by the regulatory authority in your area. This can encompass residency conditions, background verifications, financial stipulations, and adherence to zoning laws.

4.Prepare Application Materials:
Collect all essential documents and information needed for the license application, which may encompass business plans, security plans, operational procedures, financial statements, as well as property lease or ownership documents, along with any other pertinent records.

5.Submit Application:

To apply for a marijuana cultivation license, submit your application to the designated regulatory agency or department. It is crucial to adhere to all guidelines and furnish precise and comprehensive information to prevent any postponement or denial of your application.

6. Pay Application Fees:
Ensure payment of all necessary application fees linked to the licensing process. The fees may differ significantly based on the jurisdiction and the specific license applied for.

7. Undergo Inspections and Audits:
After submitting your application, it may undergo inspections and audits by regulatory bodies to verify adherence to local regulations, safety protocols, and security mandates.

8. Obtain Local Approvals:
Besides securing a state-level cultivation license, acquiring local approvals or permits from city or county authorities may also be necessary. It's advisable to consult with local government agencies to identify any further requirements.

9. Comply with Ongoing Requirements:
After obtaining your cultivation license, it is essential to adhere to continuous requirements and regulations. These may encompass regular inspections, reporting duties, meticulous record-keeping, and adherence to security measures.

10. Renewal and Compliance:
Ensure you renew your cultivation license in accordance with legal requirements and maintain compliance with all regulatory mandates to preserve your licensing status.

Consulting with legal professionals or regulatory experts who are knowledgeable about marijuana laws and regulations in your area is essential to ensure understanding and compliance with all licensing and permit requirements. Non-compliance can lead to fines, penalties, or loss of licensure, making it vital to undertake the licensing process with diligence and responsibility.

Compliance With Local Zoning Laws

Adhering to local zoning laws is essential when securing licenses and permits for marijuana cultivation. These regulations determine the permissible locations for cultivation sites, the types of facilities that can operate, and various other land use factors. The following are steps to ensure adherence to local zoning laws:

1. Research Zoning Regulations:
Research local zoning laws and regulations to understand where marijuana cultivation is permitted within your jurisdiction. Zoning laws may vary between municipalities, so it's essential to check with the appropriate local government agencies.

2. Identify Permitted Zones:
Determine which zoning districts or zones allow for marijuana cultivation facilities. This information can typically be found in the local zoning code or through inquiries with the planning or zoning department.

3. Consider Buffer Zones:
Some jurisdictions impose buffer zones around sensitive areas such as schools, parks, residential neighborhoods, or places of worship, where marijuana cultivation may be restricted or prohibited. Be aware of any buffer zone requirements and ensure compliance.

4. Check Land Use Permits:
In addition to zoning regulations, check if land use permits are required for marijuana cultivation operations in your chosen location. Land use permits may have specific requirements and conditions that must be met before approval.

5. Engage with Local Authorities:
Reach out to local planning or zoning officials to discuss your proposed marijuana cultivation operation and ensure compliance with zoning regulations. They can provide guidance on specific requirements, application processes, and any zoning variances or conditional use permits that may be needed.

6. Applying for Zoning Approval: Ensure that the selected location complies with all relevant zoning laws as mandated by local legislation before applying for zoning approvals or permits. Include all necessary information and documents to support your application.

7. Addressing Objections and Concerns:
Prepare to address any issues or objections that local businesses, associations, or residents might raise regarding your proposed cultivation operation. The zoning approval process may involve public hearings or meetings.

8. Monitoring Regulatory Changes:
For your cannabis production business, staying informed about any changes to local zoning laws and regulations is essential. As these laws can frequently change, it's vital to ensure compliance with the latest stipulations.

9. Consulting Legal Experts:
Seeking legal counsel or land use expertise from professionals who are well-versed in the local zoning laws and regulations is recommended to ensure adherence and navigate the zoning approval process effectively.

10. Maintaining compliance: Securing zoning approvals is merely the initial step. Adherence to all land use permits, zoning regulations, and any constraints or conditions established by local government bodies is mandatory. Violating zoning restrictions can lead to fines, penalties, or potentially the shutdown of the cultivation enterprise.

By carefully researching, understanding, and complying with local zoning laws and regulations, marijuana cultivators can establish operations in suitable locations and minimize potential conflicts with neighboring properties or community stakeholders.

10.2 ENSURING SECURITY AND DISCRETION

Security Measures for Cultivation Facilities

Implementing robust security measures is essential for marijuana cultivation facilities to protect against theft, diversion, and unauthorized access. Security requirements may vary depending on local regulations and the scale of the operation, but here are some common security measures for cultivation facilities:

1. Perimeter Security:
Install fencing, gates, or barriers around the perimeter of the cultivation facility to control access and deter unauthorized entry.
Use sturdy fencing materials such as chain-link or wrought iron, topped with barbed wire or razor wire for added security.
Consider incorporating landscaping elements such as thorny bushes or hedges to further deter trespassers.

2. Access Control:
Establish access control protocols to ensure only authorized personnel are granted entry.
Implement key card systems, biometric scanners, or PIN codes to regulate access to exterior doors, interior spaces, and restricted zones within the premises.
Monitor and log every entry and exit to keep track of employee and visitor movements.

3. Surveillance Cameras:
Install surveillance cameras throughout the facility to monitor activity and deter criminal behavior.
Position cameras strategically to cover critical areas such as entrances, exits, hallways, grow rooms, storage areas, and perimeter fencing.
Ensure cameras have high-resolution capabilities, night vision, and remote viewing capabilities for effective monitoring.

4. Alarm Systems:

Install alarm systems to detect unauthorized entry, intrusion, or tampering.
Use motion sensors, door/window sensors, glass break detectors, and panic buttons to trigger alarms in case of security breaches.
Connect alarm systems to a monitoring service or security company for rapid response to emergencies.

5. Physical Security Measures:
For securing doors, windows, and other points of entry, it's advisable to install robust locks, deadbolts, and to reinforce the frames.
Installing shatter-resistant glass and adding security film to windows can aid in deterring forced entry attempts.
The installation of bollards, security bars, or roll-down gates should be considered to enhance physical protection.

6. Security Lighting:
Illuminate the exterior and interior of the facility with bright lighting to deter intruders and enhance visibility for surveillance cameras.
Use motion-activated lights and timers to conserve energy and alert security personnel to potential threats.

7. Security Personnel:
Hire trained security personnel to patrol the premises, oversee surveillance systems, and manage security incidents. Confirm that security staff hold valid licenses and have received training in security measures, emergency protocols, and conflict resolution.

8. Inventory Control:
Implement strict inventory control measures to track the movement of plants, products, and supplies within the facility.
Use RFID tags, barcodes, or electronic tracking systems to monitor the cultivation process from seed to sale and prevent diversion.

9. Security policies and training:
Develop detailed security policies and procedures for employees, encompassing protocols for managing security breaches, incident reporting, and emergency response. Implement training and

awareness programs for all staff to ensure adherence to security protocols and foster a culture of safety and alertness.

10. Regular Security Audits and Reviews:
Regularly conduct security audits and reviews to evaluate the effectiveness of security protocols, pinpoint vulnerabilities, and take corrective measures. Keep abreast of emerging security threats and industry best practices to consistently enhance the security stance. Implementing a comprehensive security plan, customized to address the unique needs and risks of the cultivation facility, enables growers to secure their operations, safeguard valuable assets, and ensure adherence to regulatory standards.

Transporting and Selling Marijuana Legally

Transporting and selling marijuana legally involves navigating a complex web of regulations and requirements set forth by local, state, and sometimes federal authorities. Here's an overview of the general steps involved in legally transporting and selling marijuana, hope this helps:

1. Obtain Licenses and permits:
In order to lawfully run a marijuana business in your area, you must obtain the licenses and permissions needed. Usually, this covers the permits needed for production, distribution, retail sales, and cultivation.

Every state has a different set of rules governing marijuana enterprises, so it's critical to learn about and abide by the particular guidelines set out by your state's cannabis regulating body.

2. Respect Law Requirements:
Recognize and abide by any laws that control the sale and transportation of marijuana in your community. These rules may include specifications for product testing, security procedures, labeling and packaging.

Learn about any limitations on the sale of marijuana products, including those pertaining to distance from parks, schools and other sensitive areas.

3. Ensure Product Compliance:
Ensure that all marijuana products being transported and sold comply with quality control standards and potency limits established by regulatory agencies.

Products should be properly labeled with accurate information about potency, ingredients, and serving sizes, as well as any required warning labels or disclaimers.

4. Establish Security Protocols:
Adopt comprehensive security protocols to safeguard your inventory, workforce, and clientele. Consider the installation of surveillance cameras, alarm systems, and entry restrictions, in addition to employing security staff for on-site surveillance.

Ensure the security of transportation vehicles to deter theft or misdirection while in transit. Opt for nondescript vehicles and comply with all traffic rules and regulations.

5. Track and Trace:
Implement a comprehensive track-and-trace system to monitor the movement of marijuana products from cultivation to sale. This system helps ensure product safety and compliance with regulatory requirements.

Keep detailed records of inventory, sales, and transactions, and be prepared to provide this information to regulatory authorities upon request.

6. Train Staff:
Conduct comprehensive training for employees on adhering to regulations, following security protocols, and engaging in responsible sales practices. It is crucial that every staff member is aware of their specific duties and responsibilities to uphold legal compliance.

7. Stay Informed and Adapt:

Remain knowledgeable about regulatory changes and industry best practices, and be ready to adjust your business operations as needed. Engage with fellow industry professionals, become a member of trade associations, and attend educational events to keep abreast of the most recent advancements in the cannabis sector.

8. Maintain Transparency and Integrity:
Conducting your business with transparency and integrity is essential for fostering trust among customers, regulators, and the community. Compliance with regulations and commitment to ethical business practices are vital for securing long-term success in the cannabis industry.

By following these steps and maintaining strict compliance with regulations, you can operate a legal and successful marijuana business while contributing to the growth and legitimacy of the cannabis industry.

Risk Management Strategies

Implementing effective risk management strategies in hydroponic marijuana cultivation is essential to address various challenges and uncertainties inherent in the industry. Here is an elaboration on the risk management strategies tailored specifically for hydroponic cultivation

1. Compliance with Regulations:
Hydroponic marijuana cultivators must comply with specific regulations governing indoor cultivation methods, water usage, nutrient management, and waste disposal.
Stay informed about regulations at the local, state, and federal levels, and ensure strict adherence to all requirements to avoid legal consequences and regulatory violations.

2. Environmental Controls:

The implementation of robust environmental control systems is essential for regulating temperature, humidity, light levels, CO_2 concentration, and airflow within a cultivation facility.

Investing in top-tier climate control equipment, such as HVAC systems, dehumidifiers, and ventilation systems, is crucial for maintaining ideal growing conditions and minimizing the risk of crop failure due to environmental variations.

3. Water Management:

Proper water management is critical in hydroponic cultivation to prevent issues such as nutrient imbalances, root diseases, and waterborne pathogens.

Monitor water quality regularly, ensure adequate filtration and sterilization of water sources, and implement strict hygiene protocols to prevent contamination and maintain the health of the hydroponic system.

4. Nutrient Management:

Develop precise nutrient management protocols to provide plants with the appropriate balance of essential nutrients for optimal growth and development.

Monitor nutrient solution pH and EC levels regularly, adjust nutrient formulations as needed, and avoid over-fertilization or nutrient deficiencies that can compromise plant health and yield

5. Pest and Disease Management:

Implement integrated pest management (IPM) strategies to prevent and control pests, diseases and microbial pathogens in the hydroponic environment.

Utilize biological controls such as beneficial insects and microbial agents, as well as cultural practices, sanitation measures and organic pesticides to minimize the use of chemical pesticides and protect crop health.

6. Security Measures:

Enhance security measures to safeguard hydroponic cultivation facilities from theft, vandalism, and unauthorized access.

Install surveillance cameras, alarm systems, access controls, and perimeter fencing to deter intruders and monitor activity both inside and outside the facility.

7. Crop Monitoring and Data Analysis:

Implement automated monitoring systems and data analytics tools to track plant health, growth rates, environmental conditions, and yield metrics in real-time.

Analyze data insights to identify trends, detect anomalies, and make data-driven decisions to optimize cultivation practices, resource allocation, and crop performance.

8. Business Continuity Planning:

Develop comprehensive business continuity plans to mitigate the impact of potential risks and disruptions, such as equipment failures, power outages, natural disasters, or supply chain interruptions. Identify critical functions, establish backup systems and redundancy measures, and train staff on emergency protocols to ensure continuity of operations and minimize downtime.

9. Staff Training and Education:

Provide ongoing training and education to cultivation staff on best practices, safety protocols, and risk mitigation strategies.

Ensure that all employees are familiar with standard operating procedures (SOPs), emergency response plans, and compliance requirements to maintain a safe and compliant work environment.

10. Continuous Improvement and Adaptation:

Cultivate a culture of continuous improvement and adaptation to remain agile in response to the changing risks, challenges, and opportunities within the hydroponic cultivation industry.

Consistently assess and improve risk management strategies, drawing on lessons learned, staff feedback, and shifts in regulatory or market conditions, to bolster operational resilience and secure long-term success.

By implementing these risk management strategies tailored to hydroponic marijuana cultivation, cultivators can proactively identify, assess, and mitigate risks to optimize crop health, productivity, and profitability in a controlled indoor growing environment.

CONCLUSION

Final Thoughts On Hydroponic Marijuana Cultivation

Hydroponic marijuana cultivation is at the cutting edge of innovation within the cannabis industry, providing a groundbreaking method that enhances efficiency, yield, and quality. This technique allows growers to surpass the constraints of traditional farming and opens the door to sustainable and regulated growing practices.

The precision and flexibility of hydroponic systems allow growers to finely tune every aspect of the cultivation process, from nutrient delivery to environmental conditions. This level of control not only fosters ideal plant health and growth but also minimizes resource use and waste production, making hydroponics a sustainable choice. Moreover, hydroponic cultivation opens doors to year-round production, independent of external factors like seasonality or climate. This flexibility allows cultivators to meet market demands consistently and adapt to evolving consumer preferences.

However, success in hydroponic cultivation requires a deep understanding of plant biology, hydroponic principles, and meticulous attention to detail. It's a journey of continuous learning, experimentation, and refinement as growers strive to push the boundaries of what's possible.

Ultimately, hydroponic marijuana cultivation represents more than just a method of growing cannabis—it embodies a commitment to innovation, sustainability, and excellence. By embracing this approach, cultivators can cultivate premium-grade cannabis products that set new standards for quality, potency, and consistency in the ever-evolving cannabis landscape.

Looking Ahead: Future Trends and Development

In the future, numerous trends and advancements are set to influence the trajectory of hydroponic cannabis growth.

1. Technological Advancements:
Anticipate ongoing advancements in hydroponic technology, such as enhanced automation, precise monitoring systems, and comprehensive data analytics. These improvements will boost efficiency, productivity, and the ability to manage crops effectively.

2. Sustainable Practices:
With increasing emphasis on sustainability and environmental stewardship, it is anticipated that eco-friendly cultivation methods in hydroponics will become more widespread. Such methods are likely to include the adoption of renewable energy, water conservation strategies, and closed-loop nutrient recycling systems.

3. Genetic Research and Development:
Advances in genetic research and breeding will lead to the development of new cannabis cultivars optimized for hydroponic cultivation. These strains may exhibit enhanced growth characteristics, disease resistance, and cannabinoid profiles tailored to specific therapeutic or recreational needs.

4. Regulatory Evolution:
As legalization efforts progress and regulatory frameworks continue to evolve, there may be changes in laws and policies governing hydroponic cannabis cultivation. This could include expanded licensing opportunities, standardized testing protocols, and interstate commerce regulations.

5. Market Expansion:

As consumer demand for high-quality cannabis products grows, the market for hydroponically cultivated cannabis is expected to expand. This may create opportunities for new entrants, increased competition, and innovative product offerings catering to diverse consumer preferences.

6. Sustainability Practices Integration:
Growers are progressively adopting sustainable and regenerative farming methods, including organic certification, soil health management, and reducing carbon emissions. These practices contribute positively to the environment and simultaneously improve product quality and consumer confidence.

7. Vertical Farming and Urban Agriculture:
As urbanization increases, vertical farming and urban agriculture for hydroponic cultivation are becoming more emphasized. These methods enhance space efficiency, cut down on transportation expenses, and bring crop production nearer to urban hubs.

8. Research and Education:
Sustained investment in research, education, and professional development is key to fostering innovation and facilitating knowledge exchange in the hydroponic cultivation community. Such commitment will encourage the adoption of best practices, the refinement of cultivation techniques, and the elevation of industry standards.

9. International Expansion:
As global attitudes toward cannabis evolve and legalization efforts gain traction worldwide, there will be opportunities for expansion into new international markets for hydroponic cultivation. This may require navigating diverse regulatory landscapes and cultural considerations.

10. Health and Wellness Trends:
With growing interest in the therapeutic potential of cannabis, there will be increased demand for cultivars rich in specific cannabinoids, terpenes, and other bioactive compounds. This will drive research

into medicinal applications and personalized wellness products tailored to individual needs.

Overall, the future of hydroponic marijuana cultivation is bright, with exciting opportunities for innovation, sustainability, and growth. By staying informed, adaptable, and proactive, cultivators can position themselves to thrive in this dynamic and rapidly evolving industry landscape.

APPENDIX

GLOSSARY OF TERMS

1. Aeroponics: A hydroponic growing method that entails hanging plant roots in the air and intermittently sp**raying them with a nutrient-rich solution.**

2. Cannabinoids: Cannabis plants are comprised of chemical compounds such as THC (tetrahydrocannabinol) and CBD (cannabidiol), each with unique psychoactive and therapeutic properties.

3. EC (Electrical Conductivity): The measurement of nutrient concentration in hydroponic solutions reflects the solution's ability to conduct electricity.

4. Hydroponics: A technique for cultivating plants without soil, utilizing a water solution enriched with nutrients to supply vital nourishment directly to the roots of the plants.

5. pH: A scale that measures the acidity or alkalinity of a solution is called pH. It is critical to maintain proper pH levels in hydroponic cultivation to ensure the accurate uptake of nutrients by plants.

6. PPM (Parts Per Million): PPM, or parts per million, is a unit of measurement that expresses the concentration of a substance in a solution. In hydroponics, PPM frequently measures the concentrations of nutrients.

7. Reservoir: A reservoir in a hydroponic system serves to contain the nutrient solution, which is then pumped or circulated to the plant roots.

8. Root Zone: The area around a plant's roots where nutrient absorption occurs. In hydroponic systems, the root zone is directly exposed to the nutrient solution.

9. Substrate: The material used to support plant roots in a hydroponic system, such as rockwool, coco coir, or perlite.

10. TDS (Total Dissolved Solids): A measurement of the total concentration of dissolved solids, including minerals and nutrients, in a solution. TDS is often measured in PPM.

11. Vertical Farming: A method of cultivation that involves stacking growing layers vertically to maximize space efficiency, often used in indoor hydroponic systems.

12. Wicking System: A passive hydroponic system where plants draw up water and nutrients through a wick from a reservoir below.

13. Yield: The amount of usable plant material, such as flowers or buds, harvested from a cannabis plant. Yield is influenced by various factors including genetics, growing conditions, and cultivation techniques.

RESOURCES AND REFERENCES

1. Books:
"The Cannabis Grow Bible: The Definitive Guide to Growing Marijuana for Recreational and Medicinal Use" by Greg Green
"Hydroponic Food Production: A Definitive Guidebook for the Advanced Home Gardener and the Commercial Hydroponic Grower" by Howard M. Resh
"Marijuana Horticulture: The Indoor/Outdoor Medical Grower's Bible" by Jorge Cervantes

2. Online Resources:
Grow Weed Easy (https://www.growweedeasy.com/): An online resource with articles, guides, and forums covering all aspects of cannabis cultivation, including hydroponics.
I Love Growing Marijuana (https://www.ilovegrowingmarijuana.com/): A website offering cultivation guides, tutorials, and product reviews for both beginner and experienced growers.
Maximum Yield (https://www.maximumyield.com/): An online magazine and resource hub covering hydroponic gardening, including articles, product reviews, and industry news.

3. Research Papers:
"Nutrient Management in Recirculating Hydroponic Culture" by Lynette R. Morgan, published in HortTechnology.

Hydroponics: A Versatile System to Study Nutrient Uptake and Metabolism of Plants" by Hendrik Poorter and Urs Schmidhalter, published in Plant and Soil.

"Hydroponic Systems and Water Management in Hydroponic Cultivation" by Rafael Fernández and Dolores Rey, published in The Spanish Journal of Agricultural Research.

4. Industry Associations:

Cannabis Horticultural Association (https://www.cannabishorticulture.org/): An organization dedicated to promoting best practices in cannabis cultivation, including hydroponics.

Hydroponic Society of America (https://www.hydroponicsocietyofamerica.com/): An association focused on advancing hydroponic agriculture through education, research, and advocacy.

5. YouTube Channels:

Everest Fernandez (https://www.youtube.com/user/MrCannab1s1nside): A cannabis grower known for his informative videos on hydroponic cultivation techniques and tips.

MedGrower1 (https://www.youtube.com/user/medgrower1): A channel offering tutorials and walkthroughs of hydroponic cannabis grows, with a focus on maximizing yields and quality.

These resources provide a wealth of information on hydroponic marijuana cultivation, from beginner basics to advanced techniques and research findings. Whether you're just starting out or looking to expand your knowledge, these references can serve as valuable guides on your journey as a hydroponic cannabis cultivator.

RECOMMENDED SUPPLIERS AND BRANDS

1. General Hydroponics: Offers a wide range of hydroponic nutrients, growing systems, and supplements.

2. Botanicare: Known for its high-quality hydroponic nutrients, supplements, and growing mediums.

3. Advanced Nutrients: Provides advanced nutrient solutions and supplements specifically designed for cannabis cultivation.

4. Hydrofarm: Offers a variety of hydroponic systems, lighting solutions, and growing accessories for indoor cultivation.

5. FoxFarm: Known for its organic soil mixes, fertilizers, and supplements suitable for both hydroponic and soil-based cultivation.

6. Sunlight Supply: Provides hydroponic lighting systems, grow tents, ventilation equipment, and other indoor gardening supplies.

7. GHE (General Hydroponics Europe): Offers innovative hydroponic systems, nutrients, and additives designed for optimal plant growth.

8. Bluelab: Known for its precision pH meters, EC/TDS meters, and nutrient monitoring solutions for hydroponic growers.

9. Hortilux: Provides high-quality grow lights, including HPS (High-Pressure Sodium) and MH (Metal Halide) bulbs, suitable for hydroponic cultivation.

10. Rockwool: A popular brand of inert growing medium made from volcanic rock fibers, commonly used in hydroponic systems.

11. Vivosun: Offers affordable hydroponic equipment, including grow tents, ventilation fans, and carbon filters.

12. Growers House: An online retailer offering a wide selection of hydroponic supplies, equipment, and accessories from various brands.

When selecting suppliers and brands for hydroponic equipment and supplies, it's crucial to evaluate product quality, customer reviews, and compatibility with your cultivation setup. Comparing prices and examining customer feedback are also important steps to make informed choices for your marijuana cultivation requirements.

INDEX

A

Aeroponics, 15, 32
Air Flow, 45-47
Automated Systems, 55-58

C

Cloning, 72-76
CO_2 Levels, 38-40
Compliance with Regulations, 60-62
Crop Monitoring, 25-28
Curing Techniques, 90-92

D

Deep Water Culture (DWC), 20-22
Disease-Resistant Strains, 80-82

E

Environmental Controls, 42-44
Essential Nutrients, 30-31

G

Germination Techniques, 68-70
Glossary of Terms, 95-98

H

Harvest Timing, 86-88
High-Yield Strains, 78-79
Humidity Control, 46-48
Hydroponic Systems, 10-12

I

Indica vs. Sativa vs. Hybrid, 74-75
Integrated Pest Management (IPM) Strategies, 50-52

L

Light Cycles, 34-36
Light Requirements, 35-37

N

Nutrient Solutions, 28-30
Nutrient Adjustments, 32-33
Nutrient Requirements, 29-31

P

Pruning Techniques, 64-66

R

Recommended Suppliers and Brands, 96-98
Resources and References, 93-95
Rooting Hormones, 67-68

S

Security Measures, 60-62
Signs of Maturity, 84-85

Supplemental Lighting Techniques, 39-41
T

Temperature Control, 42-44
Transplanting Seedlings, 70-72
Trimming and Manicuring Buds, 88-90
V

Ventilation Systems, 43-45

www.ingramcontent.com/pod-product-compliance
Lightning Source LLC
Chambersburg PA
CBHW071209240526
45470CB00018B/1644